The
Endangered
Kingdom

The Wiley Science Editions

The Endangered Kingdom

The Struggle to Save America's Wildlife

ROGER L. DISILVESTRO

Wiley Science Editions
John Wiley & Sons, Inc.

NEW YORK / CHICHESTER / BRISBANE / TORONTO / SINGAPORE

For my mother and her father,
both gone from my life too soon;

he fired my interest in wildlife and wild things during
childhood conversations, on a porch beneath starry skies;
she helped me immeasurably to follow my own inclinations.

Publisher: Stephen Kippur
Editor: David Sobel
Managing Editor: Ruth Greif
Editing, Design, and Production: G&H Soho, Ltd.
Illustrator: Amy Bartlett Wright

This publication is designed to provide accurate and authoritative information
in regard to the subject matter covered. It is sold with the understanding
that the publisher is not engaged in rendering legal, accounting, or other
professional service. If legal advice or other expert assistance is required,
the services of a competent professional person should be sought. FROM A
DECLARATION OF PRINCIPLES JOINTLY ADOPTED BY A COMMITTEE OF
THE AMERICAN BAR ASSOCIATION AND A COMMITTEE OF PUBLISHERS.

Library of Congress Cataloging-in-Publication Data

DiSilvestro, Roger L.
    The endangered kingdom : the struggle to save America's wildlife /
Roger L. DiSilvestro.
        p.   cm. — (Wiley science editions)
    Bibliography: p.
    ISBN 0-471-60600-6   0-471-52822-6 (pbk)
    1. Wildlife conservation—United States. 2. Endangered species—
United States.   I. Title.   II. Series.
    QL84.2.D57   1989
    333.95′413′0973—dc19   88-38884
                                                                                 CIP

Printed in the United States of America

91   92   10   9   8   7   6   5   4   3   2

Foreword

Although many North American wildlife species have returned from the brink of extinction, and public awareness of the plight of nonhuman organisms is probably at an all-time high, the prognosis remains bleak. Habitat alteration and destruction because of human activities threaten virtually all nonhuman organisms. The only exceptions are a suite of species, such as house sparrows, starlings, and brown rats, that thrive in human-modified environments. Furthermore, even in an era of wildlife "management," overharvesting still threatens many game species.

A central and most valuable aspect of Roger DiSilvestro's book is his discussion of the role of hunters in conservation, past and present. There is little doubt that hunters have made major contributions to saving some groups of organisms—deer and waterfowl are obvious examples. Duck hunters, for instance, help support government conservation efforts when they buy the required federal waterfowl hunting stamps (duck stamps). And a private organization, Ducks Unlimited, has contributed large amounts of money to purchase critical wetland habitat. Hunters have been so important to waterfowl conservation in this century that some ornithologists think shorebird populations would be in better shape if *they* were hunted.

There is unfortunately another side to the story. As DiSilvestro points out, the hunting community has pressured managers to keep the bag limits on some ducks too high. Duck hunters also long resisted the necessary transition from lead to steel shot (to avoid decimation of flocks from lead poisoning). In general, hunters' interests have concentrated too much on maintaining game without adequate concern either for the long-term viability of their own sport or for the well-being of nongame wildlife. These priorities are often reflected in management practices. In the interest of deer hunting, for example, managers create clearings in woodlands, although clearings are inimical to many migratory bird species that thrive best in unbroken stretches of forest.

The Endangered Kingdom explores the history of the dangerously close relationship between hunters and wildlife managers. The latter group still tends to focus excessively on the needs and desires of those whose sole interest in wildlife is hunting it.

This situation is gradually beginning to change, however. The rise of the discipline of conservation biology is symptomatic of the great concern of biologists over the accelerating extinction crisis. The efforts of organizations such as the Society for Conservation Biology and Stanford University's Center for Conservation Biology are directed in part toward bridging the gap between academic scientists and those at the frontline of conservation—and that liaison is helping to expand the horizons of wildlife managers.

Recruiting such disparate groups as hunters and birders into a unified international conservation effort would be an important step in preserving civilization itself. Human beings are now using, coopting, or destroying roughly 40 percent of the food resources that all terrestrial animals depend upon. At the same time, the human population is "planning" to double its size before the middle of the next century. That prospect alone implies doom for many nonhuman organisms. Indeed, it is likely that clearing, burning, and other direct human alterations of tropical moist forests will lead to the extermination of many millions of other species and billions of genetically distinct populations before 2020.

Equally or more threatening are the indirect human assaults on natural ecosystems. Climate change due to greenhouse warming, acid deposition resulting from anthropogenic emissions of nitrogen and sulphur oxides, and ozone depletion due largely to the release of chlorofluorocarbons all will help force many populations and species to extinction. Those populations and species are working parts of ecosystems that provide services essential to sustaining civilization. In destroying them, humanity is attacking its own life-support systems.

There is, as you will see, much more to DiSilvestro's book than a fine discussion of hunting and game management, such as his compelling chapter on the slaughter of rattlesnakes. If and when conservationists can get people concerned about the fate of such creatures, then we'll have made a major step forward in the battle to save Earth's endangered living resources.

Every American with an interest in our natural heritage should be moved by *The Endangered Kingdom*. By being well informed, many diverse constituencies can come together as a major political force. They can put the sort of pressure on politicians that will be required to preserve our treasured wildlife and lessen human pressure on natural systems.

PAUL EHRLICH
Stanford University

ACKNOWLEDGMENTS

Special thanks are due to the people who read portions of my manuscript for critical review: Jack Ward Thomas, U.S. Forest Service biologist; Lovett Williams, Jr., a wildlife biologist formerly with the Florida Fresh Water Fish and Game Department; James Yoakum, a wildlife biologist with the federal Bureau of Land Management; Rollin Sparrowe, chief of the federal Office of Migratory Bird Management; Stephen Fritts, a U.S. Fish and Wildlife Service biologist; Chris Servheen, federal grizzly bear recovery coordinator; John Borneman, of the National Audubon Society; Mark A. Fraker, a biologist with Standard Alaska Production Company; George Pisani, a herpetologist at the University of Kansas; Mike Pelton, a research biologist at the University of Tennessee; Paul Robinson of Bat Conservation International, the private organization in Austin, Texas, that supplied most of the information used in the bat chapter; and George Powell, of the National Audubon Society. These people have helped me to avoid errors in fact and interpretation. I am solely responsible if any such errors remain. I also want to thank my editor at John Wiley and Sons, David Sobel, for his enthusiasm and encouragement, and my agent, Stephanie Laidman, for always having all the answers about the mysterious process of book publishing. In particular I owe a great deal of gratitude to my friend Carolyn for ungrudgingly permitting me to put into this book time that we might have spent together.

Contents

What did it all mean . . . that there should be
this beauty, so ever-varying, so soul-sufficing,
so complete, and face to face with it these
people who one and all would gladly have
exchanged it for any one of a hundred other things . . .

—Edith Wharton

Introduction

Please imagine that you are on an old wooden sailing ship, moving westward across unknown seas. You go forward into the bow, lean against the bowsprit, and peer out at the heaving blue horizon. Cold wind sweeps through your hair, the masts creak behind you as the sails strain before the wind.

And then you see it, far far ahead: a pale green line on the horizon, so thin that at first you are not sure it is actually there. The ship rolls with the waves, the horizon tilts, the masts groan.

Imagine all this as if in a dream, so that the rules governing time and distance no longer apply. Imagine, then, that you are now sailing close upon the shore, as if seconds sufficed to traverse the distant miles to the pale green horizon. You see before you deep woods and wave-swept, rocky shores. And you accept this, blending without question, as you would in a dream, the known with the unknown, the familiar with the strange.

Please imagine, then, that five centuries have slipped away from you and from that continent edging up before you. It lies untrammeled and undiscovered, touched only briefly and in passing by men in dragon-headed boats who named it Vinland and relegated it to legends for a long time only half believed. The only race to people it for any length of time has left the land unscathed, content with the provender secured by stone-age strivings. And because your understandings are like those of a dream, you perceive without knowing how that you are seeing this land as it was for the millennia and millennia before the tattooed sails of a fifteenth-century Italian explorer pierced the seaward horizon.

NEW ENGLAND

You pass island after island as you approach the mainland. The islands look like irregular fragments shed by the continent into the sea. Some bristle with pines and firs, but most are bare outcroppings of stone. Waves swill over their jagged shores, where the broken rock is covered by seaweed. The islands seem inhospitable enough, windswept and barren, but in fact thousands of large birds swarm along the edges of each one. The birds are called great auks, and they look like penguins. Their backs and wings are black, their breasts white, and they waddle upright over the shore. They are reminiscent of squat gentlemen in tuxedos. If you go ashore, they will try to flee from you, trundling away at a speed that matches your normal walking gait. If you catch one it will snap at you with its long, powerful beak. But basically the bird is defenseless. It has no enemies on the barren spits of land where it lives and so has evolved no really effective means for dealing with them. Because auks will flee from your approach they can be easily herded together by the dozens or even the hundreds and driven across the island like sheep.

The auks' wings are short and narrow, more like flippers than wings. The birds are flightless, they are clumsy and awkward on land. Sometimes they fall forward and use their wings to scoot themselves along, sledding on their rotund little bellies. Eventually they reach the sea, the element to which they have become so well adapted. In the water they take off swiftly, cutting through the swells. Some plunge below and resurface with fish in their beaks. Some merely dive under the surface and torpedo along, coming up only for occasional breaths. Their swimming is clearly in a northerly direction, and in fact they are led by unknown urges to islands much farther north. There they will nest on islands all across the north Atlantic. Now is their time of migration.

You can watch them go, little men in tuxedos swimming for distant lands. But they do not ply these seas alone. Another migrating sea creature, the Atlantic gray whale, also travels in groups along these shores. This peaceful behemoth moves gently through the water, its head emerging slowly, then tucking back under the waves, followed gradually by the massive bulk of its body. It does not breach dramatically, it does not thrash the sea into foam. It rolls quietly and inexorably north, as if time were on its side. If it were to spot a ship going by, it might even stop its effortless glide and suspend itself upright in the water. It would watch the boat with dark, enigmatic eyes. You can learn nothing easily of gray whales, even if you watch them for long hours as they pass within sight of shore.

Eventually you reach the mainland and hike across a sandy beach to the tangle of trees that forms the first obstacle between ocean winds and the deep inland forests. Twisted and shaped by the prevailing winds, the trees are low and gnarled. You pass through them to a darker wood where giant trees stand like columns in a great cathedral. Beneath the trees feed droves of wild turkeys. You see sixty, eighty of them in a flock. The flocks run as you approach. The birds glisten like polished bronze when they pass through the shafts of sunlight that sift through the trees. The woods seem alive with them, their numbers are so great.

You pass along a stream that courses swiftly in the cool shade of the trees. You hear something cracking and crunching as you search for a ford on the other side. You stop when you spot a long, slim animal lying beside the stream. It is a giant sea mink, fully half again as large as any inland mink. Its fur is coarser and redder than that of other minks, too. It gnaws at a freshwater clam. You look again at the stream and see that the bottom is paved with clams the size of dinner plates. Sunlight twinkles on water clear as glass.

Something calls from the woods on the other side of the water. It makes an odd tooting cry and howls and cackles and ends with a sound like eccentric laughter. You move quickly downstream, find a ford and cross over. When the water overruns your shoes it feels icy cold.

On the other side you listen again for the call. When you hear it you move in its direction and discover a clearing, great and flat. Perhaps trees fail to grow there because lightning has burned out the area, or perhaps the soil is too dry and sandy. In any case, it is the perfect theater for a drama that occurs there each spring. Fascinated, you watch the performers.

They are rather drab birds, mostly buffy brown, maybe grayish, with dark bars on their feathers. They look like over-stuffed chickens designed by Puritans. They are scattered all over the clearing, as far as the eye can see, each surrounded by its own little plot of ground. There each bird is engaged in some stage of an elaborate ritual in which the males cajole one another into fights. The cock begins with a short run, after which he dances in one place, stamping his feet hard enough to be heard a good ten yards. Then he stretches out his neck, holds his wings firmly against his body, and lets his long wing feathers drop to the ground. His tail fans open as if he were a peacock, and he inflates orange, featherless air sacs on either side of his neck. Then he makes a loud tooting sound and a series of cackles, followed by deep booming notes that throb through the woods for up to two miles. Now he may jump high into the air and spin completely around, then land and look at the scores of opponents with which he shares the courting ground. The spinning leap may stimulate other cocks to do the same, until the clearing is hopping with heath hens. Eventually, two males may run straight at each other, pause briefly for threats, and then charge into a spectacle that looks much like a chicken fight. Sometimes the cocks find a female feeding nearby and strut around her, but she usually ignores them.

This ritual can be seen and heard throughout the eastern woodlands. The birds number in the tens of thousands. They are so numerous around the future site of Boston that after coloniza-tion, servants will put a clause in their contracts limiting the number of times per week that they will be fed heath hens. The birds are part of the apparently endless natural resources of the new continent.

PENNSYLVANIA

The ancient forests here are not like the woodlands you remember from trips through the twentieth-century Alleghenies. The trees bear the marks of a thousand winters, the growth of a thousand summers. For centuries now they have lofted their crowns skyward, driven by hunger for sunlight. Their trunks are so wide that your outstretched arms cannot span them. But among the goliaths are younger trees, trees that grow in clearings carved by lightning fires or the falling of trees that have lived out their long lives. This mix of old and young trees makes the canopy uneven and allows sunlight to reach the floor, stimulating an undergrowth that feeds creatures you do not expect to see in these eastern woods. One is the bison. But this is not the animal that you know from Hollywood Westerns. The eastern bison is bigger and darker. Old bulls are nearly solid black. The herds are thousands strong and move through these woods from clearing to clearing. In all, some five million bison live in the eastern woodlands.

The bison share the forest with vast herds of elk. In winter, the bull elk's trumpeting is heard throughout the woods, a challenge to other bulls. The elk are so numerous that they have trampled through the trees pathways as wide as roads. At salt licks along the river that later people will call the Susquehanna they gather in herds of up to eighty animals.

The elk and the bison are hunted by wolves. On some nights, the forest is awash with their howls. They roam everywhere on the continent, from the sea through the plains to the far western mountains and beyond. Some forty thousand inhabit the woodland area that will some day be named for William Penn, one for every square mile.

Another meat eater lurks here, too, shattering many a night with a shriek that settlers will say sounds like that of a woman being murdered. It is the catamount, called in other places and other times the puma, cougar, mountain lion, panther. Like the wolf, the elk, and the bison, its numbers are immense, and you find its tracks crisscrossing woods and fields everywhere in the East. But there is another creature whose vast populations dwarf those of even these animals.

You walk through a clearing in the woods. In its midst is a moldering forest giant, its decaying trunk green with moss, crusty with lichens. The clearing is thick with new growth, warm sunlight melts over your shoulders.

Then you hear the sound of distant thunder, and the clearing dims abruptly. You believe it is a sudden storm and hasten for the shelter of the trees. The darkness deepens, and you look skyward for the clouds. Instead you see birds, infinite numbers of them speeding by, wheeling overhead layers deep. You stand to watch them pass, but they do not pass. Flock after flock goes by, blotting out the sun, wings thundering, on and on.

You watch and watch and still the flight does not end. You walk back into the woods and continue on your way. And even as you walk you hear them above you. When you reach a wide river with broad banks you sit and look skyward. The birds stream from horizon to horizon, an endless river rushing by, miles and miles of them passing on and on. A peregrine falcon bursts in upon them, and with a noise like thunder the birds rush into a compact mass to swerve away from it. The peregrine disappears, but because these birds follow the exact path of those that went before, the whole flock now swerves when it reaches the point of attack. Their colors flash in the sun, the glistening azure of their backs deepening to the rich purple of their breasts as they careen overhead.

Only evening, when the birds seek roosts in trees, puts a stop to the flight. Then the birds, all from a single flock of wild passenger pigeons, settle so densely on the branches of the forest that trees two feet in diameter are crushed to the ground. They take every available space on every branch and shrub, they land upon one another, they stack up two and three deep. Shrubs take on the appearance of swarming bee hives, for the birds are in constant motion, seeking permanent roosts. You have never seen anything like them before, nor has anyone you know. This single flock numbers more than a billion birds. The total population for the whole continent may be close to five billion. Passenger pigeons alone make up nearly half of all birds that occur at this time in the region that will be the United States. Their numbers seem indescribable, incomprehensible, and inexhaustible.

THE PRAIRIES

Fire and drought keep trees from the prairies. The flora that survive here need deep roots to save water and greenery that can be sacrificed without death to the plant. Consequently, grasses dominate the range. In some parts they grow tall enough to hide a man on horseback. In others the grasses would reach only to his stirrups.

As you walk across this wide land the grasses whisper at your feet, sweep at your fingertips. The land rolls gently away all around you. Animals that need woods to survive are relegated to life along the river bottomlands, where hardwoods grow densely. If you explore along one of these rivers at nightfall in the southern plains that one day will be called Oklahoma, you will find the roosts of wild turkeys, creatures of woods. A single roost might hold two thousand birds at night and run a quarter mile wide for a full mile along the river.

You stop atop a long rolling hill and look out over the plains, squinting against the sun. The wind waves over the grass, the prairie ripples like a green sea. You watch dark shapes that move over it—plains bison. They range in all directions as far as you can see, moving idly along, grazing as they go. No one will ever know for sure how great was their number—some will say forty million, some sixty, the biggest of them a ton on the hoof. You can walk for days and never lose sight of them, nor of the pronghorn that feed with them.

The pronghorn look like small deer, except that on their heads the bucks bear shiny black horns that end in a slight hook. Below the hook is a small prong. The horns are used mainly in battle against other pronghorns during the mating season. The animals' primary defense against other species is flight. It may be the fastest creature that ever tread the New World. It can hit speeds up to fifty miles an hour for short jaunts. When it runs it erects the white hair of its rump, broadcasting a warning to its kindred that danger is near. It may number some forty million animals throughout the plains.

On the plains you see elk, too, their many-tined antlers reaching backward over their shoulders. They are myriad, grazing on every hill and in every valley, like the bison and the pronghorn.

The hunters are here, wolves in numbers so dense that fifty or sixty of them will gather to feed on a single bison carcass. As you wander this land they, too, will rarely be out of sight. They share the bison with tens of thousands of plains grizzly bears, their tawny hair bleached nearly white by the sun.

Eventually, in your wanderings, you reach a place where the high grass stops and large mounds of overturned earth punctuate the plains. It may lie below sloping hills in a sort of prairie-valley floor, or it may be a part of a vast flat, but you recognize it immediately by the mounds. It is a prairie dog town, populated by plump, reddish rodents that run between the mounds and stand upright atop them to look at you. If you get too close to one it will throw back its head and let out a chirping whistle that alerts other prairie dogs to your presence. Get ever closer and it will drop into the burrow entrance surrounded by the mound.

You may walk all day through the prairie valley, but you will never leave sight of this town. You may walk all day again tomorrow and still not test its limits. After a week, or ten days, you may reach the end. Virtually every six paces will bring you past another mound and, until you are too near, each mound will have standing atop it at least one plump little resident that will announce your arrival. A single prairie dog town may run 100 miles wide and 250 miles long. It may hold 400 million prairie dogs.

In the town live other animals that use the burrows and that feed upon the prairie dogs. These are the black-footed ferrets, long, slim animals that resemble weasels in shape but that are honey-buff in color, with black feet, black tips to their tails, and black masks across their dark, shiny eyes. They do not weigh half as much as their prey. They move over the prairie as low to the earth as the grass, as supple as the wind. They appear above ground mostly at night. If you took a light with you to look for them, and you shined it on one, its eyes would glow green in the dark.

WEST COAST

Far up the coast, in the northwestern lands where summer days are long and winter nights all but eternal, you take a seat on an arctic shore and look out to sea. They are out there, wallowing on the waves. They could be small whales, they are so large. They could be overturned boats floating at the surface, they are so nondescript. But they are Steller's sea cows. They look something like giant walruses without tusks. Some are fully 30 feet long. Except for the whales, they are the largest sea mammals ever.

With their heads submerged the sea cows feed on seaweed. They paw at the plants with their front flippers, which are tipped with pads of thickened skin. The underside of each hooflike pad is thickly covered with half-inch bristles, good for holding on to rocks and caressing mates. Every few minutes each sea cow raises its head and breathes, blowing out air with a snort. When tired, the sea cows roll over on their backs and snooze.

At high tide they float so close to shore that you reach out and touch one. Its hide is rough, like tree bark. As you watch it float by you see clearly for the first time its immense size. You look at others in the herd, peaceful giants drifting quietly at sea, males and females, with the young kept for protection in the center of the herd.

Seemingly much later, you come to the end of your dreamy imaginings. You sit on a beach far to the south of the arctic waters, in what some day will be California. Golden bears pad over the beach, a procession of grizzlies that comes out of the gentle green hills behind you. Their paws leave deep hollows in the sand. They come one at a time, they come as females with their cubs. All are drawn by a scent on the wind, the promise of bountiful food. The sun above the quiet azure sea glows over them, everything golden, burnished by the light.

You see their destination, which lies at the edge of the gently foaming waves. Huge dark birds circle over it, black vultures with triangular white markings on the undersides of their wings. These are California condors, the largest flying birds in the world. As they soar, wind whistles over their wingtips. They too study the bears' destination, a whale that has run aground. It is like a massive shadow on the edge of the sea. One long flipper sways above the sand, animated by gusting wind. The giant vultures settle atop it, look down at the bears. Their beaks are too weak to pierce the whale's skin. They wait for the bears to finish so that they can peck at the opened flesh.

You watch from the grassy dunes that overlook the beach. The bears are brawling beside the ruined whale. You hear their bellowed threats. The bears are the color of freshly minted pennies, truly the stuff of dreams.

The sea wind is a lulling breeze. It eases away all that surrounds you. Imagine that everything you have witnessed ended long ago, in many places and many times. Imagine that you are in a world where none of these will ever be again.

TODAY

We have been pretending to see the United States as it was a few years before Columbus sailed into the seas of the New World. It is not an easy task for someone accustomed to the empty forests, plains, and skies of today's America. In fact, what we have done only hints at the diversity and quantity of animals that made up the biota of North America. We have left out many of these creatures, or have touched upon them only briefly, because we will discuss them in later chapters. Consequently, the wild America we pretended to see was a relatively pale substitute for the real thing. But even a pale substitute of something so extraordinary provides stirring images. Five billion pigeons? At least five billion prairie dogs? Fifty million bison on the Great Plains alone? Forty million pronghorn? These electrify a wildlife enthusiast's ponderings even as they provide the basis for a conservationist's list of sorrows. All that we saw has been lost.

As early as 1785 it was clear that the great auk was on its way to extinction, hunted down for food, bait, feathers, and fat. George Cartwright, for sixteen years a resident of the Labrador coast, in July of that year witnessed the slaughter of the auks on Funk Island, a nesting ground off Labrador. He wrote in his journal that if the hunting were not soon stopped, "the whole breed will be diminished to almost nothing . . . for this is now the only island they have left to breed upon; all others lying so near to the shores of Newfoundland, they are continually robbed." Yet throughout the following decades seamen continued to come to Funk Island, dragging ashore huge cauldrons and building impoundments surrounded with stone fences. Into the impoundments they herded tens of thousands of great auks and beat them to death with spiked clubs. The first killed were stacked beneath the cauldrons, now filled with water, and their fat bodies used to fuel the flames that would bring the water to a boil. Into the steaming cauldrons went thousands more. Their feathers were scalded from their bodies, slaked off the water, and bagged for shipment to the mainland. The denuded bodies were left on the shore, to be dealt with by the tide.

The slaughter was completed on June 3, 1844, when the last known of the wild great auks, a nesting pair, were killed on an island off Iceland and their single egg thrown into the sea.

The Atlantic gray whale disappeared even earlier. The last of them probably went into the whaler's rendering vats sometime in the early eighteenth century. Almost nothing is known of this whale. It vanished before science could observe it. In fact, for many years biologists doubted the descriptions that early whalers left of this species and argued that it had never existed. But recently discovered bones, the youngest of them dating to 1675, testify in melancholy silence to the Atlantic gray whale's life and death.

The sea mink, largest species of mink in the world, also perished under the onslaught of commerical hunters. Its large size made it much prized by hide hunters, who used dogs to pursue it along the rocky shores and coastal islands that it inhabited. When a mink sought refuge in a hole or crevice, the hunters either dug it out with shovels and let the dogs kill it or shot it and pulled it out with sharp-tipped iron rods. If these techniques failed, the mink were smoked out with sulfur. In the 1860s, hides were worth $8 to $10 apiece. In 1880 a hunter sold one to a buyer in Jonesport, Maine. It was the last sea mink ever seen.

The wild turkey, too, was wiped out over much of its range. Whole subspecies were lost as human settlement advanced. Connecticut saw the last of its wild turkeys in 1813, Massachusetts in 1851, Ohio in 1880, Illinois in 1903, Iowa in 1907. In some states they did survive, however, and these birds have helped in the reintroduction of wild turkeys to areas from which our ancestors extirpated them.

The heath hen was less fortunate. Abundant on the sandy scrub-oak plains of Massachusetts, Connecticut, Long Island, New York, New Jersey, and Pennsylvania, it was heavily hunted. By 1791, its numbers had dwindled so greatly on Long Island that some residents tried to enact a law to protect it. The law failed, however. By 1840 the heath hen was gone from the mainland of Massachusetts and Connecticut. By 1870 they survived only on Martha's Vineyard, off Massachusetts. These finally succumbed to disruption of their habitat in 1938. The game bird that one colonist described as too abundant to waste a shot on was gone.

The slaughter of America's wildlife went on and on, uncontrolled, unmitigated, and virtually unwept. In Pennsylvania, a huge hunt was held near Pomfret Castle on West Mahantango Creek in 1760. The hunters stood a half mile apart, forming a circle a hundred miles in circumference. They marched into the circle, constricting it and killing whatever they saw. The count was 41 cougars, 109 wolves, 112 foxes, 114 bobcats, 18 bears, 2 elk, 98 deer, 111 eastern bison, 3 fishers, an otter, 12 wolverines, 3 beaver, and some 500 smaller animals.

The last Pennsylvania circle hunt was held in 1849 in Center County. With such prolonged slaughter, the hunted species had to dwindle. By the end of the nineteenth century, the cougar had been wiped out throughout the Northeast. A tiny population in southern Florida is all that remains of the eastern cougar today. Of the perhaps two million wolves that once roamed the region now called the lower forty-eight states, only twelve hundred remain, all but a few in northern Minnesota. The last elk in Pennsylvania was killed in 1867. Elk found in that state today are transplants from the West. In 1825 a hunter near Tygart's River in West Virginia located an eastern bison cow and calf. They were the first bison seen in the East in ten years. The hunter killed both, bringing to an end the history of the eastern bison.

And so it went. Even the land itself was mangled. By 1900, 96 percent of the eastern woodlands had been logged at least once. The prairies were plowed for agriculture, and the Southwest was so badly overgrazed by livestock that we now have desert lands where early Spanish settlers found grass growing as high as their horses' bellies. In the plains, the bottomland hardwood forests where turkeys roosted have been mostly flooded by dam projects. Even the grizzlies that badgered the Lewis and Clark expedition at so many river bends have lost 99 percent of their numbers to unrestricted hunting, government predator control programs, and loss of habitat. Today in the West no more than a thousand survive, and even their fragile existence is increasingly jeopardized by human encroachment. Federal and private poisoning programs have decimated the vast prairie dog towns and probably have contributed to the near extinction of the black-footed ferret as well. Fewer than twenty ferrets are known to survive today, all in captivity.

The rest that we saw also succumbed to one extent or another. The California condor survives only in zoos. The plains bison, whose vast herds shook the earth, were hunted mercilessly for meat and hides and reduced to only five hundred animals by the turn of the century. Today, they have recovered some of their losses, but the ten thousand that remain are all fenced in parks and preserves. The seven-ton sea cows, discovered in 1741 by the Vitus Bering expedition, were hunted into extinction only twenty-seven years later. Georg Wilhelm Steller, after whom the species was named, described how sailors killed one of the animals by boating out to it and sinking a large, sharp hook into its flesh. To the hook was attached a rope, and men on shore dragged the sea cow to land, where it was beaten with clubs and stabbed with knives. Meat was actually cut from its back while it was still alive. Steller also described how a male sea cow tried to defend a hooked female, willingly followed her to land, and remained with her butchered body for at least three days.

Perhaps the clearest measure of the profligacy of early America is the demise of the passenger pigeon. We have imagined its billions as best we could. But can we even begin to imagine how those billions could be slaughtered in the nineteenth century down to the last bird? Is it useful to know that hunters could fire shotguns into migrating flocks and kill seventy birds with one shot, more than a thousand before breakfast? Or that hunters killed thousands at night by burning pots of asphyxiating sulfur in roosts and by setting fire to trees to frighten flightless squabs out of nests? Does it help to know that one shooting club used 50,000 passenger pigeons in one week for targets? Or that in 1878, hunters killed a billion pigeons near Petoskey, Michigan, from a single nesting site that was forty miles long and three to ten miles wide? Is it salient that through the use of nets, a single professional hunter could kill nearly five thousand pigeons in just one day? Does it tell us something about the breadth of the slaughter to know that in 1822 one family in Chautauqua County, New York, near Lake Erie, killed four thousand pigeons just for the feathers? Or that boxcar after boxcar of the birds were shipped East for sale in city markets and that one New York dealer sold eighteen thousand pigeons per day?

A flock of passenger pigeons that once flew over Ontario was estimated to hold 3.7 billion birds. But even such numbers could not withstand the unbridled killing. By 1896 only a quarter million survived. They gathered near Bowling Green, Ohio, not far from Mammoth Cave. Hunters were notified by telegraph. It was subsequently estimated that only five thousand birds escaped. That was the last big hunt.

In Pike County, Ohio, on March 24, 1900, a young boy shot the last wild passenger pigeon of certain identity. It was not too far from the place where naturalist John James Audubon, eighty-seven years before, had watched a flock of more than a billion pigeons fly over the Ohio River.

It is said that from a distance the passenger pigeons' cooing sounded like the ringing of bells. We will never know it, never hear it for ourselves, because the twentieth century has inherited from the nineteenth a ravaged land. Nevertheless, we still have among us some few fragments of our lost wildlife heritage. How these creatures have survived, who has helped them stave off extinction, and what needs to be done to protect them are the subjects of the chapters ahead.

Part One

Game Animals

CHAPTER 1

North
American
Deer

Doubtless almost everyone who calls North America home knows the deer. The animal is certainly familiar to those who fraternize with woods, plains, or mountains. New England hikers may see them briefly, short flashes as the deer bound off into the forest, leaves crunching beneath darting hooves. Campers in Shenandoah National Park in Virginia may find their tents invaded by deer searching for food. Strollers among the rolling hills that grace the western end of the Golden Gate Bridge may turn from a hazy view of San Francisco's skyline to find a half-dozen deer looking down upon them from a nearby hill. Even those with barely a passing acquaintance with the outdoors sometimes encounter deer. At dusk and after dark motorists may see them feeding by the roadsides along Interstate 80 in Pennsylvania or along one of the many arteries that empty into New York City. Travelers may see them from trains gliding across the western plains. Thanks to *Bambi*, the animal is even a part of mass culture. The term *Bambi* has itself entered the popular vocabulary: Hunters use the expression *Bambi lover* to connote someone who opposes deer hunting, usually, the hunters presume, because of an unrealistic understanding of and an emotional attraction to deer. The epithet itself is a measure of the importance of deer to hunters and to those who serve hunters' needs. More people in more states hunt deer than any other type of big game, spending millions of dollars on equipment, travel, food, and lodging. In some states, rural towns close their schools on the opening day of deer season so students can join the hunt. The roots of this intimacy with the deer can be traced far back into history.

Between forty million and fifty million deer of three different types browsed and grazed across North America when Columbus first moored off Caribbean shores. The white-tailed deer ranged throughout the woodlands of all states and parts of southern Canada, though its occurrence in California, Nevada, and Utah was limited. The mule deer lived in the high plains and the western mountains from Mexico to southwest Canada. The black-tailed deer, a breed of mule deer, lived along the Pacific Coast from northern California to southeast Alaska.

These animals were the primary source of meat and hides for the three million Native Americans who lived within their range. Tools were made from their bones. The Native Americans killed an estimated five million deer yearly, but the deer population survived without serious decline. Deer were plentiful along the Eastern Seaboard when the first European colonists arrived. Four hundred French soldiers exploring Wisconsin early in the eighteenth century found enough deer near Green Bay to feed themselves and a thousand Indian allies for several days. A hundred years later a U.S. Army captain traveling through Texas reported seeing thousands of deer daily in herds of up to two hundred animals. It was the same throughout the Great Plains and the Pacific Northwest, the same wherever Europeans had not yet arrived.

THE YEARS OF DECLINE

From the beginning of settlement, deer were vital to the European colonists. They were the primary source of meat, and their skins were made into clothing. They were important in trade, too. Deer hides were purchased from Native Americans and exported by the thousands to the Old World. Because Native Americans were able to trade hides for guns, and guns could be translated into tribal political power, as many deer as possible were taken. In places where coin and currency were scarce, hides were sometimes used as a medium of exchange among the colonists. *Buck* as slang for money stems from this period.

By 1646, barely a generation after colonization began in the Northeast, deer numbers had been noticeably reduced in the New England colonies. That year, Rhode Island passed the colonies' first closed season on the animals. By 1705 five other colonies had done the same, but trade in deer hides continued. From 1698 to 1715, Virginia and the Carolinas exported no less than 1.1 million hides. Nearly 4 million hides were exported from Charleston, South Carolina, during a 24-year period that ended in 1765. Trade at this level occurred all along the Eastern Seaboard. By the beginning of the nineteenth century, deer numbers probably had been halved east of the Mississippi. At about that time deer hunting abated temporarily in the East. Native Americans had been pushed out, and the settlers themselves were moving west. Deer numbers surged as abandoned farms reverted to the types of browse plants upon which deer feed.

Meanwhile, the hunt shifted West, aided by a new sign of human progress. On May 10, 1869, the Union Pacific met the Central Pacific in Promontory, Utah, and the last stake was driven into the transcontinental railroad. Spanned by the trail of the iron horse, the wild lands of the West were isolated no more. Hunters could easily reach western game, and meat and hides could be shipped quickly to eastern markets. The repeating rifle was invented at about this time, facilitating the rapid killing of game.

Although the states enacted new laws for wildlife protection, the hunting of deer for meat and hides reached a new peak in the second half of the nineteenth century because the laws were weak and poorly enforced. Open seasons might run for six months with no limit on the number of deer a hunter could kill. In Wisconsin, for example, the first closed deer season, enacted in 1851, was only five months long. In summer 1855, a single hunter near LaPointe killed fifty-five deer in six weeks. During hard winters, such as the winter of 1856 to 1857, whole herds of deer that bogged down in deep snow were killed with blows to the head. No game wardens were hired in Wisconsin until 1887. Not until 1891 was a statewide ban enacted against jacklighting—the use of bright lights to blind deer at night so that they stand still and become easy targets. That same year the open season was still two months long. Selling deer meat was legal in Wisconsin until 1900.

Tales of vast killings of deer came from all over the nation and its territories. On the Pacific Coast, a single market hunter might bring down five or six hundred deer in one season. A lone hunter in the 1870s killed more than a hundred deer in a single stand in the mountains of Colorado. In Montana's Judith Basin, three hide-hunters in the late 1860s killed fifteen hundred deer in fewer than six weeks. Thirty thousand to forty thousand deer were reportedly shipped from southern Oregon every year late in the 1800s. Hides shipped to Omaha, St. Paul, Chicago, New York, Philadelphia, and other market towns sold for twenty-five cents to a dollar each, meat for as little as two cents a pound and rarely more than a quarter.

By the end of the nineteenth century, deer had been exterminated throughout much of their range. With the exception of Maine, deer had been virtually wiped out in New England. In New York they were limited mainly to the Adirondacks. They were gone from New Jersey and practically extinct throughout the Midwest from Ohio to Nebraska and in the southern portions of the Great Lakes states. Few deer remained in the West and the South. As the twentieth century loomed, the nation's deer population dropped to only 2 percent of its pristine size. Several states enacted tougher laws to protect deer and other wildlife, but even sport hunters resisted them. For example, hunters opposed laws that banned the use of jacklights and dogs, the two most deadly means for hunting deer. In New York State, Adirondack hunters did not want season limits reduced and opposed regulations that would prohibit shooting deer for use as food in camps during prolonged hunts. With such resistance to restrictions, it seemed unlikely that deer anywhere would long survive.

The slaughter might have gone on had it not been for a coterie of wealthy eastern sport hunters who feared the extinction of deer and other game animals killed for the meat markets—wild sheep, elk, moose, pronghorn, bison, and more. Among these hunters was Theodore Roosevelt, then only in his twenties and not long returned from his ranch in the Dakotas. In 1887 he helped organize the Boone and Crockett Club, dedicated to promoting strict laws for wildlife protection. Roosevelt knew firsthand the destruction that had overtaken the nation's wildlife. In 1890 he hunted in the northern end of the divide between the Little Missouri and the Powder rivers. Two decades before, bison had crowded this range. Roosevelt found the plains littered with their innumerable skulls and bones, but saw not one living specimen. He and others—such as George Bird Grinnell, editor of *Forest and Stream* magazine and founder in 1887 of the first National Audubon Society; John Muir, founder in 1892 of the Sierra Club; and William Hornaday, president of the New York Zoological Park and a volatile spokesman for wildlife—through their writings and legislative activities provided the philosophical and legal underpinnings of what came to be called the conservation movement.

By 1900, resistance to game laws was diminishing if only because the scarcity of game had made market hunting unprofitable. In many cases, former poachers and market hunters became game wardens, charged with enforcing the new wildlife laws. This era was a boon for deer. Because healthy does usually have twin fawns every year and may even have triplets, minimal protection can help deer populations rebound rapidly. They did so with such unexpected speed in the opening decades of this century that in many areas the growth proved very nearly disastrous both for the deer and the land that supported them.

RECOVERY BEGINS

In the opening decades of the twentieth century the professionals charged with protecting the nation's wildlife knew little about the needs of wild animals. They were trained primarily in fields such as range (grassland) management or forestry. In addition, little research had been done on deer and other game species, so their biological requirements were poorly understood. Consequently, land managers did not know how best to protect game animals. Their efforts to increase vanishing deer populations led to a series of errors from which some regions have yet to recover.

Perhaps the most notorious catastrophe of this era occurred along the north rim of the Grand Canyon in a part of Arizona known as the Kaibab Plateau. In 1906, during his second term as president, Theodore Roosevelt declared a thousand square miles of the plateau to be the Grand Canyon National Game Preserve. The preserve, about half of which was forested, could probably have held about fifteen thousand deer but was home to only about three thousand. It was managed by the Forest Service, which banned all hunting there and initiated an intensive predator control program. Eventually, some six hundred mountain lions, one of the primary predators of deer in the West, were killed. Perhaps more important, logging cut back the forests. This made the Kaibab a better feeding ground for deer, boosting their reproductive rate.

The Kaibab deer population skyrocketed. By 1912 the herd had more than trebled to ten thousand. Ten years later it was nearing twenty thousand and clearly in trouble. The deer had outstripped the Kaibab's ability to feed them. Their numbers needed to be reduced. Foresters sounded the first warnings of disaster in 1918 but were ignored by the deer managers, who were accustomed to believing that the deer population needed to be increased. Moreover, although managers at the time believed that livestock that grazed on the Kaibab in some cases should be removed from deer range to reduce competition for available forage, it had not occurred to them that deer, like livestock, could overgraze their habitat.

By the early 1920s most of the best deer food plants were depleted. Even less desirable foods, such as white fir, were being browsed too heavily. Famished deer were beginning to highline the Kaibab, devouring all palatable browse as far up as they could reach. Within a few more years, wildlife managers would know that such a browse line was a sign of serious overpopulation, but that understanding was only beginning to dawn in the 1920s.

By 1924 the Kaibab herd was hovering near thirty thousand, twice the estimated number of deer the area could be expected to maintain successfully. The Department of Agriculture—the federal agency that administered the Forest Service—called for a special hunt on the Kaibab in an attempt to reduce the herd by 50 percent. The governor of Arizona saw this as a usurpation of states' rights and opposed the hunt in court. While state and federal attorneys fought, the last of the forage was decimated. Thousands of deer starved to death during the severe winters of the 1920s. By 1931, the thirty thousand had become twenty thousand, and by 1939, that number was halved. Some parts of the Kaibab were so severely damaged that they could feed only 10 percent of the deer they could have carried before overgrazing.

Despite the clear lessons to be learned from the Kaibab, game specialists elsewhere doubted that deer management was at fault. The Kaibab, some argued, was surrounded by desert and the Grand Canyon. It was like an island on which deer were trapped. These special conditions had caused the collapse of the Kaibab herd.

Soaring deer populations in other parts of the country gave the lie to this viewpoint. In Black Canyon, a part of the Gila National Forest in western New Mexico, regulations similar to those of the Kaibab had been initiated by Aldo Leopold. Leopold in 1933 would write *Wildlife Management*, the first book to systematize the basic techniques and tenants of wildlife conservation. He had arrived in New Mexico in 1909, twenty-two years old and armed with a forestry degree from Yale University. Over the next ten years he had helped establish within the Gila National Forest several game refuges in which hunting was banned. Such protected refuges were then popular among game managers, who presumed that deer killed by hunters on lands surrounding the refuges would be replaced by emigrating refuge animals. To ensure that refuge deer were safe, Leopold sought to eliminate the predatory animals that fed upon them. He turned against predators the very weapons that only a decade or two before had been used against the deer themselves. In *The Pine Cone*, a local conservation publication he had founded, he encouraged the federal government to trap, shoot, or poison all animals that prey upon deer. "The sportsmen and the stockmen—one-third the population and one-half the wealth of New Mexico—demand the eradication of lions, wolves, coyotes, and bobcats," he wrote in 1919. "Game protection makes the killing of varmints necessary."

By 1927, deer at densities of roughly forty per square mile had highlined even the less desirable food plants of Black Canyon and its surroundings. An internal argument among New Mexico's wildlife managers as to whether deer were overpopulated, as well as state laws that limited deer hunting to bucks only, prevented decisive actions to reduce the herd. The area remained chronically overpopulated through the 1940s. Overgrazing so thoroughly reduced the land's ability to support deer that once a population decline began it continued into the 1970s, when deer numbers were down some 70 percent and still slowly falling.

Problems with overpopulation occurred in the north, too, particularly in states where forests had been cut for timber at the turn of the century. At first the removal of the trees allowed more sunlight to reach the forest floor, where it stimulated the growth of browse plants important to deer. The increase in food boosted the deer population dramatically. As the forests grew back, however, trees crowded out the browse plants. The enlarged deer herds then faced famine. Deer in some counties in northern Wisconsin were so weak from hunger that they could be captured by a man on foot.

By this time, Aldo Leopold was working at the University of Wisconsin and becoming increasingly involved in the state's conservation politics. His ideas about deer management had changed drastically after his experiences in New Mexico. He realized that the bucks-only hunting season held in Wisconsin was not removing enough deer to prevent the animals from overbrowsing their habitat. He therefore began urging the state to hold an antlerless deer season, one in which hunters would be permitted to kill does and even bucks too young to grow antlers. This was seen as a solution to deer overpopulation because antlerless hunts usually cause deer numbers to decline while bucks-only hunting permits them to increase. Leopold also encouraged the state to protect its few remaining predators, primarily bobcats, coyotes, and a handful of wolves, because he had come to the conclusion that predators could limit deer populations more efficiently and reliably than could human hunters.

The residents of peak deer-hunting counties joined hunters to fight Leopold on both points. While hunters were opposed to any program that might leave them with less game, local residents feared loss of revenue if the number of hunters visiting their towns declined following a reduction in deer numbers. The battle mounted through the late 1930s and into the 1940s. In the opening years of World War II, one hunter wrote an item for a local newspaper in which he compared wolves and other predators to the Nazis and Japanese. He then castigated Leopold for his support of the animals.

Those opposed to deer reduction insisted that if the deer were starving, the state wildlife agency should feed them. The state did provide winter feed from 1934 to 1954, but the feeding did not save the deer from famine. Most biologists opposed the artificial feeding, arguing that it only heightened the problem of overpopulation by keeping deer at swollen numbers. They went unheeded in the face of hunter pressure.

Many of the opinions that hunters brandished in the face of Leopold and his allies had, ironically, been taught them by early wildlife managers. Leopold himself only a decade or two before had stressed the importance of exterminating predators and hunting only bucks. The main problem for Leopold in the 1930s and 1940s was that no public relations campaigns had been launched to teach hunters that the conditions surrounding deer management had changed. Leopold belatedly attempted such a campaign by leading tours of woodland counties where deer were starving by the thousands. Participants saw deer too weak to resist when picked up. Some died just from the stress of being handled.

By 1943 Leopold had swayed public opinion far enough to win from the state wildlife-management agency a hunting season that permitted the taking of does. The hunt that year reduced the state's half million deer only 10 percent. Leopold, convinced that the herd needed to be reduced 50 percent, predicted that at the current rate of decline deer overpopulation would persist at least until 1949. Nevertheless, public outcry against the killing of does stifled further attempts to reduce the herd. Does were not hunted again for six years, and opposition to deer reduction and doe hunting continued through the 1960s. In 1970, the number of deer lost to starvation nearly equaled the seventy thousand killed by hunters.

Similar problems arose elsewhere. In some states where deer had been almost exterminated they were introduced from areas with large populations. In Pennsylvania, where deer had neared extinction in 1905, some twelve hundred deer were stocked between 1906 and 1924. Under protection of a bucks-only hunting law, the Pennsylvania herd stood at about a million deer by 1927, and signs of overpopulation were apparent. In 1928 the state initiated an antlerless hunting season. Long opposed by hunters, the season came too late. Thousands of fawns died of starvation during the two following winters. By 1931, 800,000 deer were left. More antlerless deer seasons followed, but in 1938 Pennsylvania still held a half-million deer on depleted forest habitat.

Many of the problems that complicated early efforts at deer management have survived to the present day. Wildlife agencies are still pressured to provide hay and grain for starving deer on depleted habitat. Antlerless seasons in overpopulated areas are still vehemently opposed by many hunters, especially when the hunters themselves have not observed any starving deer. However, it is clear to knowledgeable professionals that herd reductions should begin long before deer have so seriously damaged their habitat that starvation sets in. Such damage takes decades to mend and reduces for many years the number of deer the land can carry. In the long run, early reductions permit the sustained survival of larger deer herds. Research shows, however, that as a rule hunters want to see a lot of deer when they hunt. They prefer overpopulated areas with bucks-only hunting, even though they cannot shoot most of the deer they see. Ironically, antlerless deer seasons take hunting pressure off bucks, permitting more bucks to grow to larger, trophy size.

DEER MANAGEMENT TODAY

Deer are found in all forty-nine mainland states. Two types, the key deer of Florida and the Columbian white-tailed deer of the Pacific Northwest, are on the federal endangered species list. The deer is the most popular big-game animal in North America and for many hunters the only big game they will ever shoot or even see. Hunters kill about a half-million mule and black-tailed deer each year and about two million whitetails. Widespread interest in deer on the part of hunters and other wildlife enthusiasts makes management of the animals one of the most important state wildlife-conservation issues.

Deer management is directed by state wildlife agencies, usually called fish and game departments, departments of wildlife conservation, or departments of natural resources. Most wildlife agencies are administered by a board of directors, often called a wildlife commission, whose members are appointed by the governor. The commission hires the wildlife-department director, who is the individual responsible for the professional staff that runs the department. State wildlife agencies usually have several divisions, with responsibility for deer management falling to the game division or its equivalent. The game division provides the commission with the information it needs to set the time and length of hunting seasons and the season bag limits.

Most states allow three basic types of deer hunting—modern firearms, archery, and primitive firearms. Hunting with bow and arrow and with primitive firearms—muzzle-loading, single-shot rifles such as the cap and ball used in the nineteenth century—is far less efficient than hunting with modern firearms, so seasons involving use of these weapons may run two or three times as long as modern-firearms seasons.

State game divisions keep track of how hunting affects deer by studying the animals in the field and by collecting data on deer killed by hunters. Data collection varies from state to state, but the three basic tools are mail-in forms, questionnaires with telephone follow-up, and check stations.

Mail-in forms are issued to hunters who buy deer permits. They are returned to the game division by successful hunters, who fill in blanks showing when and where they shot their deer. One problem with this system is that hunters do not always send in their forms. In Pennsylvania, for example, where more than a million hunters kill up to 150,000 white-tailed deer yearly, about 40 percent of successful hunters do not report their kills.

Check stations are sites to which hunters are required to bring their deer. A check station may be a gas station or hunting supply store. In some states, check-station owners are paid by the state for each deer processed at their site. The stations are often manned by wildlife-management students who have volunteered for the program. They weigh the deer, record its sex, and remove a tooth or one lower jaw. By examining tooth condition, biologists can determine each deer's age and consequently the age structure of the state's deer herd. The age structure provides clues about how the deer should be managed.

Most wildlife agencies have divided their states into deer management units, areas in which resident deer are treated as a single herd or population. By tracking data from check stations and/or mail-in forms, combined with deer population surveys and studies of habitat condition, wildlife managers can tell which herds need to be reduced by antlerless seasons and which increased by bucks-only seasons.

Wildlife commissions are not entirely free to rely upon their professional staff's expertise and best judgment in setting seasons. They must also base their decisions upon public comments given at hearings held yearly before the deer season is set. It is at these public hearings that political pressures are most obviously applied to the wildlife commissions. And although most wildlife managers have abandoned the idea that deer conservation depends on hunting only bucks, many hunters have not. Consequently, deer biologists in every region of the country report the squelching of antlerless seasons by hunters and hunter groups that oppose deer reductions even though needed to prevent long-term damage to habitat.

Biologists often lose in confrontations with hunters primarily because hunters indirectly control wildlife-management purse strings. Deer conservation is bankrolled primarily by the sale of hunting licenses and permits. Additional funds come from the Federal Aid in Wildlife Restoration Act, signed into law by Franklin Roosevelt on September 2, 1937. Usually called the Pittman–Robertson Act after the men who ushered it through Congress—Nevada senator Key Pittman and Virginia representative A. Willis Robertson—the law put a tax on hunting equipment and earmarked the funds raised for wildlife restoration work. Today the P–R program, as it is usually called, includes an 11 percent tax on firearms, ammunition, and archery equipment and a 10 percent tax on handguns. The money is apportioned to the states based on the number of hunting licenses each state issues and the land area of each state. States receiving the funds must match every three federal dollars with one of their own. Since its inception, the P–R program has provided the states with one and a half billion dollars for use in wildlife-habitat acquisition and management, research, and hunter education.

Because hunters constitute the main source of P–R funds, the money is used almost entirely for programs designed to benefit game animals, such as deer, ducks, and geese. This arrangement gives hunters the dominant voice in shaping the management policies of all state wildlife agencies. Leopold himself questioned this system, suggesting in 1931 that obligating wildlife departments to hunters was one of the greatest obstacles to balanced conservation programs. The system will not change, however, unless a funding system is enacted that broadens support for wildlife management beyond the hunting community. Meanwhile, hunters remain the deer's most supportive allies. Without the taxes hunters are required to pay under the P–R program, deer might long ago have become extinct or at best rare.

Deer recovery since the 1930s is often touted as one of wildlife management's major successes. Wildlife biologists measure the extent of this success by the resurgence in deer numbers since the lows of the early 1900s—12 million white-tailed deer and 5½ million mule deer nationwide. Management of the deer certainly cannot be characterized as a failure. However, at least one biologist has suggested that wildlife management may be taking more credit than it deserves for the augmentation of deer numbers. For example, mule deer numbers climbed steeply in the 1950s and early 1960s, then declined into the mid–1970s before stabilizing. This, wrote Guy Connolly in *Mule and Black-tailed Deer of North America*, "reveals how little control biologists and managers have over the deer they purport to manage. Just as they were powerless to halt the decline, the biologists and managers now are unable to show in any scientifically acceptable way that improved management put the herds on the road to recovery. Nor can they assure the public that mule deer are now secure." Nevertheless, few conservationists would fail to agree that wildlife managers play a major role in keeping deer problems from turning into crises.

DEER PROBLEMS

In some states, illegal hunting for table meat and for black-market sale of venison to restaurants accounts for as many deer as the legal hunting season. Thousands of deer are also killed yearly by motor vehicles. But although illegal hunting and roadkills may jeopardize some local populations and reduce hunting opportunities, they do not threaten deer survival nationwide. Wildlife managers account for these losses when setting seasons and planning management programs. The major widespread threat is habitat loss to human development. Habitat loss cuts out the very ground from beneath the deer, leaving them fatally bereft.

Mule deer habitat has been squeezed more and more tightly over the past two decades as people have rushed to live and build along the eastern front of the Rocky Mountains. In many areas, when deer come down out of the mountains to return to traditional lowland winter feeding grounds, they find not forage but condominiums and houses. The residents of western states who see deer starving to death literally in their own backyards are observing the biological trade-off they have made in converting wildlands into family homes.

Agriculture also squeezes deer from natural habitat. Mule deer feed extensively in agricultural meadows during spring and fall, where they must compete with cattle for forage. Deer move out of areas in which cattle are grazed in moderate to heavy numbers, and cattle destroy the growing trees among which deer seek shelter.

Overgrazing of rangeland by cattle is a serious problem throughout the West. Vast parts of Arizona and New Mexico have been converted from lush grasslands to parched desert by more than a century of livestock overgrazing. Much of the overgrazed land is owned by the federal government and leased to ranchers. Politically powerful, the ranchers continue to overstock the range without fear of being penalized for damages. Wildlife such as deer cannot compete effectively against cattle and sheep because livestock eat not only the plants deer need but also those that deer find unpalatable.

Utah, Colorado, and Montana have spent thousands of dollars to feed starving deer during recent hard winters. Local residents usually support the feeding programs and may even demand that state wildlife agencies feed deer. Consequently, the agencies view the feeding programs as good public relations. Not only are urban residents pleased to see deer fed, but the programs help keep deer from depleting ranchers' haystacks. However, biologists are well aware that the programs are basically unsound because they concentrate deer at the feeding sites. This can lead to overgrazing of the area surrounding the feeding site, compounding the problem of food scarcity. And, most importantly, because artificial feeding programs can benefit only a small portion of the deer in any given area, the number saved is biologically insignificant. During the unusually harsh winter of 1983–84, the Wyoming

Game and Fish Department refused to provide feed for deer. In Utah, however, expensive feeding programs were undertaken. Both states lost roughly 70 percent of their deer in hard-hit areas.

Artificial feeding will not prevent deer population declines in the West because the declines are caused by loss of habitat to housing, industry, and agriculture. Deer biologists there are seeking ways to protect deer feeding grounds and the migratory pathways that link lower winter grounds with spring and summer feeding areas in the mountains. Deer declines could prove ominous for other wildlife species by causing biologists to ignore the other species while setting management priorities that favor deer and those who hunt them. Richard J. Mackie, a Montana Department of Fish and Game biologist who has studied mule deer for a quarter century, wrote in *Restoring America's Wildlife* that habitat "of highest potential and greatest importance may have to be rigidly protected and closely managed, perhaps by excluding other species and land uses, if reasonably abundant mule deer populations and recreational opportunities are to be maintained."

In the East, white-tailed deer managers face similar habitat losses. Changing forestry practices are causing deer declines in many widespread woodland areas. The intensive cutting of woodlands in eastern Maine for the pulp industry, coupled with increasing urban expansion around Portland and other communities in the southern part of the state, has contributed to what deer managers believe is a 25 percent drop in deer numbers statewide. In 1983, for the first time in a century and a half of controlled hunting, Maine initiated a bucks-only season for the heavily hunted southern portion of the state. The new restrictions have helped the deer population recover, but development is continuing at a rapid pace. In some areas in southern Maine, habitat is being fragmented, isolating deer into areas that hunters cannot enter. Maine's deer study leader, Jerry Lavigne, is concerned that these areas will suffer from overpopulation in the years ahead.

The seriousness of habitat loss varies from place to place. One of the most urgent deer problems is in the Florida Keys, the chain of islands that runs westward from southern Florida for 150 miles into the Gulf of Mexico. Sixteenth-century Spanish explorers who rowed ashore onto the white-beached Keys found an abundance of deer. They were diminutive creatures, miniature versions of mainland whitetails shrunk by generations of evolution in adaptation to a limited island food base. Adults stood little more than two feet tall at the shoulder. The tiny fawns could virtually stand in the palm of a human hand.

In the era when seamen still traveled under sail, ships stopped occasionally in the Keys to shoot the deer for meat. Little note was made of key deer natural history. Even the extent of the deer's original range was left unrecorded. Settlement of the Keys after 1900 led to the cutting of wooded areas for agricultural development and probably boosted deer numbers by increasing the food supply. However, the deer were soon being relentlessly hunted as pests. When wanton destruction of the deer became a national issue late in the 1930s, the state instituted a hunting ban—which local residents ignored. Key deer were tracked down with dogs, and forest fires were set to drive them into the open, where they were slaughtered. The population was whittled to only fifty deer before a special warden with both federal and state law-enforcement authority was stationed on the Keys in 1951 to guard the deer. The state also initiated a study of the deer that year. A 6,000-acre national wildlife refuge was established for them in 1957. By this time the key deer were becoming a cause célèbre and something of a tourist attraction. Hunting abated, and the key deer in 1967 was listed by the federal government as an endangered species. It and the Columbian white-tailed deer of the Pacific Northwest, which has been increasing dramatically, are the only deer in the United States that are included on the federal endangered species list.

In the mid–1970s, the key deer population reached a recent peak of some 400 animals. About 250 lived in the heart of the refuge on Big Pine Key. Perhaps another 150 were scattered over a score of other Keys. But in the following decade a housing boom swept across Big Pine Key as the island's human population doubled. Pine woods and mangrove swamps vital to the deer succumbed to development. The refuge, too small by itself to sustain the deer for long, has been transformed into an island besieged by condominiums and highways. Burgeoning road traffic is killing increasing numbers of deer. About 20 percent of all fawns drown each year in drainage ditches dug for mosquito control. Only 250 deer remain, and the number is still declining.

Thickets, woods, and open scrub still sprawl over parts of Big Pine Key, but much of it is slated for development. Biologists predict that deer numbers will sink to the tenuous levels of the 1950s if conversion of pinelands to condominiums does not stop. A halt to development seems unlikely, though, since local residents increasingly regard the miniature bucks and does as pests that destroy gardens and cause traffic accidents.

Far from the mangrove swamps of the Florida Keys, another breed of deer is being shunted toward disaster. This is the Sitka black-tailed deer, a creature of southeast Alaska's chill and foggy islands and jagged mainland, where the land rises steeply through deep forests to alpine tundra and ice-capped mountain peaks. It lives at the northwestern edge of deer range, where conditions barely meet its needs. Because its subsistence is marginal, any changes in the woodlands and tundra can jeopardize its delicately balanced existence.

Eighty percent of this land, fully 17 million acres, is held by the federal government in the Tongass National Forest. The Tongass includes the continent's largest remaining tract of virgin forest, the type of habitat without which the Sitka deer cannot survive. The forest's oldest trees—spruce and hemlock—measure their lives in centuries. They first took root a full generation before the Normans conquered England. They were still easing their crowns skyward in the seventeenth century, when colonial tradesmen where cutting down equally ancient fir trees in Maine to provide masts for British ships. They stood as ragged dark shadows in the silent fog when eighteenth-century Russian ships first charted a course among southeast Alaska's myriad islands. They survived the logging that decimated virtually all of the 850 million acres of virgin forest that once covered North America.

Traditional wisdom teaches that mature forests are poor habitat for deer. According to this axiom, the crowns of the trees form an even, closed canopy over the forest floor, shutting out sunlight and stifling growth of the plants that deer eat. This idea stems from studies of secondary forest growth—the woodlands, about a century old, that replaced the virgin forests cut down throughout most of the United States in the eighteenth and nineteenth centuries. Studies of truly mature, old-growth forests—those never cut—are revealing an entirely different picture. The canopies of virgin forests such as the Tongass are not even. Great holes are chopped into them when the ancient giants fall, killed by disease, insects, age, or some other factor. The openings let in sunlight, stimulating the growth of deer forage. In the Tongass, old growth benefits deer in yet another way. The canopy, though broken, helps shield the ground during times of deep snow. Consequently, in months when the snow lies deep deer forsake the alpine tundra, where they find the best forage, in favor of the old-growth forest. The old growth is their only refuge from the snows that cover the alpine tundra for all but three to five months of the year.

The deer may soon lose their winter refuge if the U.S. Forest Service goes through with plans outlined by a congressional act in 1980 to cut down most of the Tongass old growth at the rate of 20,000 acres a year for the next century. Once the forest is cut, three centuries will have to pass before its like will appear again. But the trees will not be given three centuries. The Forest Service plans to cut them again within the next 125 years.

Biologists who have studied the Sitka blacktail say the cut will decimate the deer. Where loggers have already clearcut the Tongass, snow piles up nearly twice as deep as it does in the old-growth areas. Even during moderate winters, recent clearcuts are buried too deep for use by deer. In the old growth, however, snow build-up is slow enough even in the worst of winters to reduce the length of time during which starvation conditions prevailed.

Even where forest land has begun to recover from clearcutting, sprouting with the shrubbery that is the best type of deer forage, the old-growth woodlands remain the most important winter habitat for southeast Alaskan deer. Deer in winter generally use old-growth forests six times as much as they use second-growth woodlands.

Of all the land held in the Tongass National Forest, no more than a half million acres is old growth. The U.S. Forest Service plans to cut down all but 160,000 acres of these ancient trees. The service is doing this partly because, under federal law, national forests are managed principally for timber production. Wildlife is a secondary consideration. Nevertheless, the cutting might be avoided in any other national forest by setting aside the old growth as a federally protected wilderness area in which no permanent human encroachment is permitted. For the Tongass, this possibility is limited by the Alaska National Interest Lands Conservation Act of 1980, which in settling the fate of Alaskan lands held by the federal government mandated that the Tongass be cut fast enough to provide a minimum annual harvest of 4.5 billion board feet of timber per decade. This provision was a giveaway to the politically powerful timber industry. The Tongass lands that have been set aside as wilderness under the Alaska Lands Act are mostly barren rock and ice.

To meet the Alaska Lands Act quota, the Forest Service must cut the best of the Tongass timberland, the old growth. To finance the cut, the Alaska Lands Act ordered the federal government to provide the Forest Service with $40 million yearly for a Tongass logging budget. The operation has provided private lumber companies with cheap timber, but it has been a financial disaster for the federal government. From 1982 to 1987 the Forest Service spent $234 million to facilitate sale of Tongass timber, but the service estimates that total income from the timber sale is less than $3 million.

Fifty percent of the old growth is slated for cutting within the next century. The trees carved out of the Tongass will be turned into pulp for paper production. Much of the pulp will go to Japan. Logged areas will lose up to three-quarters of their Sitka black-tailed deer. Other animals will be affected, too. Ten thousand bald eagles of breeding age live in the Tongass, and the logging will eliminate 90 percent of their perching and nesting sites. The nation's largest concentration of grizzly bears is in southeast Alaska, where they den during winter in the old growth.

The fate of the Sitka black-tailed deer and these other creatures hinges on passage of a proposed congressional bill that, because of the tremendous loss in taxpayer dollars that Tongass logging represents, would put an end to the 4.5 billion board-foot mandate. It would also help preserve one of the few places in North America that has eluded human progress, one of the few that provides a chance to see the continent as it was seen by the earliest explorers.

Despite the problems faced by the Sitka black-tailed deer, key deer, and deer populations throughout the country, it seems unthinkable that survival of the deer will ever again be seriously jeopardized. The proficiency of modern wildlife managers should bar forever a general collapse of deer numbers. It may well be, however—if development goes unchecked—that two generations hence the deer will have been driven once again from vast portions of its range.

C H A P T E R 2

The Wild
Turkey

The turkey is a poorly appreciated specimen of North American wildlife. It is likely that many Americans do not even know that turkeys live wild all across the continent below southern Canada. These wild creatures are not the bulky barnyard turkeys that occupy the imagination and please the palate every November, those nearly flightless birds that, when not depicted lifeless on a platter, are usually shown puffed up into the fan-tailed breeding posture of the males. The wild bird is sleeker, longer legged, too highstrung and flighty to survive long in captivity. On foot it can outrun a man and quite possibly a dog. On the wing it is one of the swiftest of North American birds, capable of reaching 50 miles per hour for up to a mile. It is mantled in feathers that in sunlight glisten like bronze.

The wild turkey is a New World native. The first people to see it were the anonymous wanderers who some fifteen thousand or more years ago drifted out of Asia across a now-vanished land bridge that connected Alaska to Siberia. Wild turkeys were doubtless among the animals that supplied the explorers with food. Remains left by later peoples certainly suggest this. Judging from the number of bones they left behind, the people who some five thousand years ago inhabited what today is called Kentucky fed upon turkeys more than any other animal except deer. Fifteen hundred years ago the residents of the Kansas City, Missouri, area apparently ate only deer and raccoons more than turkeys. The Native Americans of more recent times developed an intimacy with the wild turkey that they shared with few other animals, since the turkey and dog were the only animals they domesticated. The Aztecs kept thousands of turkeys in pens, so many that Montezuma, when he governed the Aztecs, could exact from his subjects a levy of a thousand turkeys daily for the feeding of his private zoo and his household.

More northerly peoples also kept domestic turkeys. The Conquistadors reported that the tribes throughout the lands now called Arizona and New Mexico kept turkeys in pens. DeSoto, moving north out of Florida in 1540, found captive-reared turkeys in Georgia. In North Carolina, a Cherokee chief gave him a gift of seven hundred turkeys.

Although the Native Americans of the Southeast ate turkey meat, many of the peoples of the Southwest kept the birds only for the feathers, which were woven into cloaks, and for the bones, which could be shaped into tools. Some southwestern tribes avoided the meat because they thought eating it would make them cowardly. They did not seem to mind sharing it with others, however. The Spanish Conquistadors, who tortured and killed many Native Americans while scouring the Southwest for cities made of gold, were often given turkey meat by the Native Americans they met. Once Cortes was even told that he and his men could have a meal of turkey, bread, and cherries if they were humans, or five Natives if they were gods. Perhaps the tribes of the Southwest were hoping to weaken the will of the fierce Spaniards by diluting their bravery with turkey meat. If so, it must soon have become clear that the old beliefs that guarded the native peoples were faltering, that the old certainties were on the wane.

THE PERIOD OF EUROPEAN SETTLEMENT

In 1502 Christopher Columbus made his fourth voyage to the Americas. While sailing the Caribbean he explored some islands off Central America, one of which he named Guanaja. In August he landed at Point Caxinas in what today is called Belize. The Native Americans he met there provided him with food, including the meat of a bird he dubbed *gallina de la tierra*. It is not clear just what bird he was referring to, but some scholars think it was the turkey and that Columbus was, therefore, the first European to see the species. They offer three arguments in support of their view. First, *gallina de la tierra* had never been used before for any New World bird. Second, the term was soon reserved by the Spanish strictly for the turkey. And last, the Cuban word for turkey is *guanajo*, perhaps derived from the island that Columbus had christened Guanaja and near which he first saw the *gallinas de la tierra*.

Regardless, the turkey would soon be well known even in Europe. Early in the sixteenth century, Cortes and other Spanish explorers brought some of the domesticated birds to Spain. Within two or three decades domestic turkeys had spread to Italy, Germany, France, and England. By 1573 turkeys were common table fare among English farmers. Turkey meat was so much in demand that settlers traveling to the Americas sometimes brought European-bred turkeys back to the New World. The bird's place in the barnyards of the world was thus assured. In the wild, however, its survival was increasingly uncertain as Europeans flooded American shores.

The history of the turkey during European colonialism parallels that of other wildlife used for trade or food. When the first sailing vessel nudged an American beach, more than ten million wild turkeys ranged across the northern continent, from New England west across the Mississippi and southwest across the Rio Grande. The governor of Plymouth, in preparation for the first Thanksgiving, sent four men into the woods for game. In one day they returned with enough turkeys to feed the town for nearly a week. A traveler in colonial Maryland might count a hundred or more in a single forest glade as the birds rushed on wing and foot to feed upon grasshoppers. An early riser wending a path along the edge of a North Carolina swamp might watch myriad flocks depart at sunrise, some flocks five hundred strong. Armed with a single-shot muzzleloader, a woodsman in Kentucky could find enough turkeys to shoot thirty, forty a day with ease. In southern Ohio, flocks of a hundred to five hundred birds coursed the woods. Farther west, where the woodlands withered to plains, turkeys in the thousands fed over the prairie, returning at night to roost among the trees that traced the pathways of rivers and creeks.

Nowhere were turkeys more numerous than in the Oklahoma and Texas country. Gen. Phil Sheridan, camping in Oklahoma's Ellis County three years after the Civil War, reported that the trees were black with turkeys. In the late 1870s, wanderers in the Oklahoma hills found flocks of three thousand birds and roosts a quarter mile wide and a mile long.

Tracking turkeys in snow was considered good sport. Dogs were also used to hunt them. Circle hunts, in which a group of hunters encompassed an area of woods and then slowly constricted the circle by moving inward, brought down as many as five hundred turkeys at a time in early nineteenth-century Ohio. Hunting at night was also effective because at sunset turkey flocks would retreat to their roosts, stands of trees in which they would perch until daylight. Hunters would shoot the birds off the lower branches first, then gradually work upward. The shooting did not disturb the roosting turkeys as long as they were not hit by falling bodies. Using this method, two hunters in Ohio killed 130 turkeys during four or five days of hunting.

Under intense hunting pressure the wild turkey faded first from the East, vanishing from Connecticut in 1813, from Vermont in 1842, from Massachusetts in 1851. Travelers in the West, heedless of the consequences of unrestrained hunting across the Mississippi, continued the slaughter. In 1880, Gen. George Crook and some men in his command killed more than a hundred turkeys while hunting one evening and one morning near Fort Apache, Arizona. In Oklahoma, whole wagonloads of turkeys were slain in a single night. So many turkeys were shot for market that birds of up to 20 pounds could be bought for only a few pennies. As the 1800s came to a close, the turkey disappeared from Ohio, Nebraska, Kansas, and the Great Lakes states. Early in the 1900s it was killed off in Iowa, Illinois, and Indiana. Two or three decades later only a few hundred survived in Oklahoma.

The loss has often been blamed on logging, forest fires, and agriculture, but Florida turkey biologist Lovett Williams, Jr., in *The Book of the Wild Turkey*, placed the blame on the uncontrolled hunting of the nineteenth century. "In early America," he wrote in 1981,

> the gun came before the axe, and excessive exploitation depleted turkey populations well in advance of the more material signs of civilization, just as it did in the cases of the bison and the panther. . . . I will always believe . . . that the wild turkey could have survived [the inroads of human settlement] in good shape if it had not been overhunted.

The Wildlife Managers Step In

Early wildlife managers knew little of the turkey's needs. They did not even know with any certainty the type of habitat that turkeys required. A booklet about wild turkeys published in 1924 by the Field Museum of Natural History reported that turkeys lived primarily in swamps. Armed with such information, conservationists presumed that swamps were therefore good turkey habitat. They had yet to learn that in the 1920s the birds were restricted largely to swamps only because these wetlands were among the last places unpillaged by agriculture and logging. Several more years would elapse before biologists established that the turkey's prime habitat lay in hardwood forests, where the birds could feed upon acorns and other fruits. It was not until 1943 that the first comprehensive book on the species, *The Wild Turkey in Virginia: Its Status, Life History, and Management*, was published. By then, however, little hope remained for the turkey's survival. The first thorough research on wild turkeys, initiated in the 1930s, had been arrested by the Second World War. In 1944, the last of New York State's turkeys were extinguished.

This period was among the darkest for the wild turkey. Populations were at low ebb—the bird was extinct in eighteen of the thirty-nine states in which it was originally found and threatened with extinction in five. The turkey's range was drastically reduced—the birds had been wiped out of nearly 90 percent of their habitat east of the Mississippi and nearly 70 percent of their range in the West. And wildlife biologists were beginning their second decade of misguided management.

Early managers sought to recover the wild turkey in much the way they had sought to recover deer, using techniques that satisfied human logic but fell short of meeting the needs of nature. One mistake was an emphasis on predator control. J. Stokley Ligon, a federal field biologist and colleague of Aldo Leopold, made these comments in a 1946 government publication about wild turkeys in New Mexico:

> The universal seriousness of predation to turkey restoration, it is felt, fully justifies the prominent consideration given the subject. Turkeys apparently are an irresistible attraction to coyotes, and the question as to whether the white man or the coyote has had the most adverse effect on the Mountain turkey is debatable.

He added that unless the government consistently and intensively worked to kill off coyotes, bobcats, and eagles, sportsmen could not be certain of having surplus turkeys to hunt.

Ligon, whose avid support of predator control had contributed to the management debacle of early deer conservation, was instrumental in getting thousands of dollars diverted into killing coyotes and other predators. These funds might have been better used on research into turkey biology. Wildlife managers today almost universally agree that predator control for the protection of wild turkeys is a waste of limited management funds and scarce personnel. This conviction jelled in the 1950s as increasingly sophisticated studies were conducted on the turkey. Lengthy research on predation in New Mexico showed that Ligon's assertions could not bear scrutiny. Biologists found no significant increases in turkey populations in areas where predators had been controlled. They also found that predators scarcely preyed upon birds more than two weeks old. Biologists in the Southeast confirmed that adult turkeys in their region were seldom killed by predators. In Montana and Wyoming, biologists found that birds trapped in the wild and released in new habitat were able to establish healthy populations without any predator control. Research conducted as recently as 1987 shows that even where predation is a major cause of death it may have only a minimal effect on turkey populations: In northern Missouri predators account for 55 percent of the hen turkeys lost each year, yet the area has one of the state's highest turkey populations. Contemporary wildlife management calls for use of predator control only in local cases in which large numbers of birds or eggs are being lost to predators. Abandonment of excessive predator control helps save resources for more useful purposes.

The other, and probably bigger, mistake made by early managers was an attempt to restock depleted turkey range with birds reared on state wildlife agency game farms. These birds usually had at least some domestic turkey ancestry, making them plumper than wild turkeys and difficult to frighten. Even after they were released into the wild, they failed to flee from danger. Consequently, captive-reared birds rarely survived more than a few weeks after release, even in good habitat, and programs based on them universally ended in failure. The failure was compounded in many areas because the released birds carried diseases that reduced the already constricted wild populations.

Attempts to rear pure wild birds in captivity also failed. Excessive hunting had made them extremely wary and nervous. When approached by predators of any kind, including humans, they would promptly flee, usually by running. If hard pressed they would fly. Too high-strung to tolerate confinement, they could not survive long enough in captivity to produce the number of birds needed for stocking programs.

Some states persisted until the late 1950s in stocking domestic and semi-domestic birds, particularly in eastern states where turkeys had been wiped out. However, the real solution to turkey depletion had been discovered as early as the 1930s. This was the relocation of wild birds from areas with healthy populations to areas where the turkey had been wiped out. Relocations were first attempted in the West, where good turkey populations survived in some pockets, and met with unexpected success. In 1935, fifteen wild turkeys from New Mexico were released in Wyoming. Two decades later Wyoming had an estimated ten thousand. South Dakota followed suit in 1949, releasing twenty-nine turkeys from New Mexico and Colorado. By 1960 the state had as many as seven thousand birds.

The traps used to capture wild birds in the West did not work well with eastern turkeys. Consequently, relocation programs were stymied in the East until 1948, when researchers at Swan Lake National Wildlife Refuge in Missouri developed a capture technique that used rockets to fire nets over flocks of turkeys attracted to the nets by bait. Since then, use of cannon nets has become a well-established technique for capturing wild turkeys. Another more recently developed technique has also become popular—use of drugged bait, such as corn that has been treated with narcotics. The birds eat the bait and pass out. Because they eat until the drugs take effect, overdosing can be a problem. Consequently, this technique requires close monitoring by biologists.

These techniques permitted relocations of eastern birds on a large scale and made the stocking of wild-caught turkeys a common practice. Without these techniques the nation's turkey population would not number today some two million to three million birds, many of them in states outside their original range, including Montana, California, and Hawaii.

AT THE TWILIGHT
OF THE TWENTIETH CENTURY

Hunting, increasingly well managed since the turn of the century, no longer threatens the turkey's survival. Seasons are carefully restricted, with limits placed on the numbers and sex of birds that hunters can shoot. Some states allow turkey hunting only in the fall, after the nesting season. Others permit a spring season, when hunters hiding in camouflage garb can use a call that imitates the yelp of a hen turkey to attract males within range of the gun.

As with the management of other wildlife hunted for sport, turkey conservation is funded primarily by the Pittman–Robertson tax on hunting equipment and by the sale of state hunting licenses. This arrangement permits the hunter's voice to speak more loudly than any other in setting turkey-management policy. Consequently, some state wildlife agencies are forced to continue using management practices that their professional staffs have long recognized as futile wastes of funds. For example, hunters in some areas demand the stocking of game-farm turkeys. Although years of experience have shown that game-farm birds quickly die in the wild, many hunters remain convinced that the releases will increase the wild turkey population and their own chances of shooting a bird. Ironically, the releases pose a threat both to wild populations and hunting because diseases carried by game-farm birds can lead to local declines in wild turkey numbers.

These problems are minor, affecting primarily local or regional turkey populations. A more serious obstacle to the survival of the wild turkey is the nationwide ebbing of hardwood forests. These woodlands, dominated by trees such as oak and hickory, are crucial to wild turkeys. They provide the birds with acorns and other nutlike fruits, vital fall and winter foods. But not one region of the nation has stanched the loss of hardwoods. Hardwoods in the South are being cleared over vast acreages to make way for the pine plantations favored by lumbermen who want trees of all one type, all one age, ready for cutting on schedule. Thus, wildlife dependent on native southern forests are increasingly in decline. Throughout the United States, thousands of acres of hardwood stands along rivers and streams have been drowned behind dams and beneath reservoirs. In the East the peaked roofs of newly built housing developments and the glittering edifices of modern shopping malls have replaced the boundless woodlands that once must have made new settlers feel infinitesimally small.

Nothing jeopardizes wildlife as seriously as loss of homeland. Backed by an ever-watchful and technically advanced system of wildlife management, the wild turkey seems safe from all the dangers it has survived, and yet the species is already waning in some parts of the East. The fate of the wild turkey, like that of any wildlife species faced with the inexorability of human progress, remains as distantly unknown today as it was during the time of slaughter a hundred years ago.

CHAPTER 3

The
Pronghorn

The pronghorn was more truly a creature of the Old West than
the bison, cougar, or elk. These three, associated today with the
Rockies and the plains, ranged far into the East before they were
checked by urban civilization. Their bones lie beneath the soil of
Pennsylvania, New York, and the Virginias. But the pronghorn
lived only in the West, where they grazed the boundless prairies
that rolled and rolled to the far horizons or eked out a living in
desert fastnesses where rivers dried to powder each year. They
avoided the woodlands, except when driven there for shelter
during winter storms, adhering in most seasons to open country
where they could see enemies coming from miles away.

The first Europeans to see pronghorn were the Spanish Conquistadors. They took little note of them, perhaps because the
pronghorn were too small to provide much meat and wore hides
too thin to be more than marginally useful. Coronado must have
seen them in 1540 when he traveled through New Mexico and
into central Kansas, but the only indication is mentioned in his
journals of fleet "goats" seen in Arizona and of "deer pied with
white" in New Mexico. These can only have been pronghorn.

The creature remained obscure until the early nineteenth century, when Lewis and Clark filed reports on the "goats or antelope" they saw late in 1804. With tawny backs, white bellies, and

hooked black horns, pronghorn do indeed resemble goats or antelope. But they are neither, though even today they usually are called pronghorn antelope. They belong in a family entirely their own, native only to the plains and deserts of North America. They are the only animals that annually shed the outer layers of their horns. The black sheaths, with their downward-curving tips and small prongs near the midpoint, fall off in November in both males and females but are replaced before the following spring.

Lewis and Clark first spotted the pronghorn near the site of Greenwood, South Dakota, and killed a male. About two weeks later, on September 17, 1804, Meriwether Lewis grew tired of cruising the Missouri, rowed ashore, and wandered into the plains somewhere west of the White River. In his journal he described what he saw—a prairie dog town three miles long, surrounded by a plain so lush and closely cropped that he compared it to a bowling green; immense herds of elk and deer grazing across the plain; so many bison that he guessed he saw not less than three thousand within a single view; and, dappling the prairies with their myriad numbers, the pronghorn.

"My object," he wrote, "was if possible to kill a female Antelope having already procured a male. . . ." Through the morning he followed several herds, never getting close enough for a shot because the pronghorn were "extreemly shye and watchfull." He watched one fleeing herd disappear at a run behind a ridge, then reappear three miles distant so quickly that he doubted they were the same animals. "But," he wrote, "my doubts soon vanished when I beheld the rapidity of their flight along the ridge before me it appeared reather the rappid flight of birds than the motion of quadrupeds." Captain Lewis doubtless never knew that he was watching the fastest four-footed animal in the New World, capable of hitting 50, perhaps even 60 miles per hour for short sprints. He also could not have known that a man would need eight-power binoculars to match the pronghorn's eyesight. With the equipment Lewis and Clark had at hand, the pronghorn could be easily approached, shot, and killed only when it floundered while traversing a river. The animal was challenging game even for the wolves that roamed in great numbers over the plains. Yet the pronghorn barely survived the nineteenth century.

A CENTURY OF DECLINE

The waning of the pronghorn, like the waning of the bison, echoes the fall of the Wild West. At the start of the century the prairies were unbroken, the plains unfenced. The pronghorn, as little known as the West itself, was of interest primarily to Native Americans. They sometimes constructed long barriers of brush or branches that led into crude corrals. Then they would herd scores of pronghorn toward the barriers, which the animals would follow into the corral. There the pronghorn would be slaughtered. Native Americans also tricked pronghorn into coming close enough to a concealed hunter for an easy shot by hanging a piece of cloth or hide where a pronghorn could see it flapping in the wind. Drawn by curiosity, the pronghorn would investigate the object while the hunter hiding nearby took aim. White hunters in the nineteenth century learned this trick and used it to supply meat for the markets.

Europeans pursued the pronghorn for sport as early as 1540, when a hunt was held in the Mexican state of Hidalgo in honor of the local viceroy. Participants killed six hundred deer driven by Native Americans. The take included deerlike creatures, unknown in Spain, that were said not only to run but to fly. Presumably, pronghorn.

Except for this story, and the scant comments of the Coronado expedition, little mention is made of the pronghorn in the early writings. Until after the Civil War, vast, unsettled grasslands insulated the animal from the human inroads that had decimated eastern wildlife.

While that natural barrier persisted, visitors to the West saw the pronghorn in the same teeming numbers that characterized so many other wildlife populations in untouched America. Unfortunately, few early observers diverted themselves long enough from describing bison, elk, and wolves to leave a clear picture of the dimunitive pronghorn in a pristine age. But a few records provide images that quicken the pulse and inspire the envy of anyone whose eyes ache for the vanished sights of a long-lost wilderness America.

The pronghorn was said to rival, if not surpass, the bison in numbers. A hiker along Kansas's Salina River in 1873 saw bands of hundreds of pronghorn "as far as the eye could survey." George Bird Grinnell, a naturalist who traveled the West with George Custer and his cavalry not long before the Last Stand, wrote that the troops were surrounded by pronghorn for day after day as they traveled the prairie. Near North Park, Colorado, he saw so many pronghorn that they seemed to cover the land. The Ute Indians in 1868 killed 4,400 pronghorn in North Park during one circle hunt. Bands of two thousand or three thousand pronghorn were common on the California plains. Many fell to miners' guns during the gold rush.

In winter, the numbers seemed even more overwhelming because pronghorn from miles around would gather in sheltered areas or to migrate to winter feeding grounds. As late as 1896 a wanderer along Montana's Milk River could find forty thousand pronghorn herded together in the coulees, seeking protection from the great blizzard that struck that December. During the winter of 1868–69 a passenger on the newly opened railroad that traveled between Denver and Cheyenne would have witnessed a sight no modern visitor can ever hope to see—two million pronghorn gathered in the foothills into a single herd 10 to 12 miles long and 100 to 200 yards wide. The animals were so densely packed, like a living blanket on the range, that they changed the color of the countryside. Wagonloads of the animals were killed throughout the winter and brought to Denver. Carcasses sold for about six cents apiece.

As late as the 1870s, the pronghorn population may have exceeded forty million. Within three decades the animal had all but vanished. The U.S. Biological Survey reported in 1908 that the entire U.S. population did not exceed seventeen thousand. Gone too were the vast numbers of bison, deer, and elk that had enchanted Captain Meriwether Lewis in 1804. Gone too were the unbroken plains and the untouched prairie.

LIVESTOCK AND THE PLOW

Market hunting played a major role in the collapse of the pronghorn, especially after the railroads opened the West and the repeating rifle made shooting easier. But settlement itself had an equal hand in the debacle. Settlers swarmed into the West in the second half of the nineteenth century, encouraged by federal homesteading programs begun in 1862 and by the ease of railroad travel. Using plows to turn the prairies upside down, the sodbusters that settled the Dakotas, Nebraska, Kansas, and parts of other western states destroyed the pronghorn's sustenance over hundreds of thousands of square miles. The destruction of the pronghorn was under way.

Further west, in Arizona, New Mexico, Colorado, Montana, and elsewhere, much of the land was too dry or rugged for farming. But the grasslands there were superlative for livestock. The Spanish had discovered this three centuries before. And from the Spanish the American rancher learned a new kind of livestock husbandry, the grazing of the open range. In the early 1870s five million cattle were feeding on western lands. Within two decades the number topped twenty-five million. They roamed wild on the unfenced prairies, fending for themselves. Long horned and rangy, they could be dangerous to people who approached too closely on foot. They were rounded up once a year, and selected numbers were driven to market. The rest of the time they lived unmolested on the open range, most of which was public land administered by the federal government. Because ranchers had no vested interest in protecting the land, and because each rancher knew that his rivals would put out as many cattle as possible, the land became overridden. Far more cattle were grazed than the land could support. The deep grasses that had nourished millions of wild grazers vanished into the craws of domestic stock that ate vegetation to the ground before moving on. The cattle also tended to concentrate around streams and other water sources. As a result, over large parts of the Southwest they destroyed stream-bank vegetation and caused the collapse of the banks themselves. Widening of the streams made the water run shallow and raised the water temperature. This killed off fish and changed the dynamics of water flow, causing many streams to dry up completely.

The damage the cattle did has outlived the era of the open range. Where grasses once whisked the bellies of the Conquistadors' horses, today there are deserts. Some 225 million acres of western lands, as many acres as comprised the original thirteen colonies, are still turning to desert, and twice that much are threatened with desertification. Most of the remaining rangelands are in poor condition, grasslands overrun with brush that grew in the wake of overgrazing.

The enduring wastelands that overgrazing created were once the lush feeding grounds of the pronghorn. Widespread loss of habitat and massive slaughter for the meat market posed nearly unsurmountable obstacles for pronghorn survival. In Colorado the pronghorn population fell from twenty-five thousand in 1898 to two thousand within a decade. In Yellowstone National Park, hunters reportedly killed thousands yearly from 1872 to 1883. The military took over administration of the park in 1886, but no punishment was meted out for illegal hunting until 1894. By 1908 not more than two thousand pronghorn remained in Yellowstone. Within five years that number dwindled to only five hundred. In 1913, naturalists in Texas and Wyoming—men within whose lifetimes the pronghorn had sunk from millions to thousands—were reporting that the animal would soon be extinct within their states. William T. Hornaday, director of the New York Zoological Society, wrote that year in his book, *Our Vanishing Wildlife*, "The prong-horned antelope, unique and wonderful, will be one of *the first species of North American big game to become totally extinct.* We may see this come to pass within twenty years." As he wrote that, the Wool-Growers Association in Montana, advocates of sheepmen, was hiring a Washington lobbyist to stop a proposal in Congress that would have set aside

15 square miles of prairie around Snow Creek, Montana, as a national pronghorn preserve.

PROTECTION BEGINS

Several developments at the turn of the century helped stem the decline of the pronghorn. In many parts of the West the land was too dry or infertile for farming. The homesteads the government had given to settlers of these marginal areas were too small to support a single family, so they were abandoned. Slowly, the broken land reverted to natural plant growth, and the sodbusters' fences rotted back into the earth. The pronghorn thus won back parts of its native habitat.

The grazing practices of western ranchers also changed. Their own excesses conspired with bad weather to cause tremendous losses of livestock in the late 1880s. For the first time, ranchers began to worry about the state of the land. Early in the next century they started to push for federal legislation to control grazing. The land continued to be overgrazed, and a federal grazing law was not passed until 1934, but the birth of a rudimentary land ethic nevertheless helped diminish the number of cattle grazed, reducing competition with pronghorn.

And, finally, pronghorn hunting came under control as the conservation movement brought an end to market hunting. Passage of the Pittman–Robertson Act in 1937 helped provide funds for research on pronghorn habitat and behavior. One benefit of the research was improved relations with ranchers, many of whom had opposed pronghorn protection because they thought the animals ate the same grasses as cattle and spread livestock diseases such as brucellosis. Neither assumption was proved correct. By 1944 the Arizona Game and Fish Commission could report that ranchers were willing to cooperate with efforts to increase pronghorn numbers.

One of the most important steps in pronghorn recovery was undertaken in 1937 by the New Mexico Department of Fish and Game. This was the relocation of pronghorn from areas where they were abundant to areas where they had been wiped out. The pronghorn were captured by herding them toward fencelike barriers and into corrals, much as Native Americans had done centuries before.

The technique worked well, and other states soon adopted trap and transfer programs. Arizona surveyed its herds and outlined a relocation project in 1944. Colorado relocated a thousand pronghorn in one decade, while Texas relocated five thousand from 1939 to 1975. In some cases relocated pronghorn simply returned to the areas in which they were captured, but despite such setbacks the programs were highly successful. Pronghorn numbers have exceeded a million throughout the animal's original range since the early 1980s, more than a half million of these in Wyoming alone. The only big-game animal that outnumbers them in the United States is the deer, and hunters annually kill more pronghorn than any other big game except deer.

AND TOMORROW?

The recovery that wildlife managers initiated has not yet peaked. Pronghorn numbers are still increasing, though no state wildlife agency seems able to explain why. Diminished use of rangelands by cattle during a period of depressed beef prices may be a factor.

Many of the grazing problems that destroyed pronghorn habitat in the nineteenth century have barely been alleviated. Nearly 4.5 million cattle, sheep, and goats are grazed on federal lands in the West. In some areas they are joined by large numbers of wild horses and burros. Government surveys and independent studies have shown that 60 to 80 percent of the federal lands are in unsatisfactory condition because of overgrazing. The figure in 1936 was 84 percent, so improvement since passage of the federal grazing law is slight. These lands, administered by the Bureau of Land Management, are leased to ranchers at cut rates. The charge for grazing one adult cow on federal lands for one month hovers at about $1.35. Meanwhile, the charge on private lands is nearly $7. Some ranchers have even sublet the federal lands they lease, charging other ranchers fees that exceed $12. Conservationists, concerned that public grasslands are being destroyed by livestock at give-away prices, believe the federal charge should be raised to the level of the private market. However, though the ranchers who lease federal lands are few in number—they make up about 8 percent of all western ranchers and raise only 2 percent of the nation's beef—they are politically powerful. Consequently, Congress has refused to change the fee system. Were the number of livestock grazed on public lands reduced and grazing fees raised to increase the funds available for range management, then programs could be undertaken to restore grasslands that have been devastated for a century or more. This restoration could lead to further increases in pronghorn numbers and to better forage for both wildlife and cattle.

Despite the condition of the range, the pronghorn is secure as long as the vagaries of human population growth and of economics do not place further limits on its habitat. Ironically, though, the very success of pronghorn recovery can stimulate new conflicts, particularly among ranchers who see the growing number of pronghorn as competitors for livestock forage. One Wyoming rancher made national news in the winter of 1984 when he erected a fence to lock two thousand pronghorn out of a 22,000-acre area called Red Rim, where he grazed cattle. Red Rim is a traditional pronghorn winter feeding ground because high winds sweep it clear of the deep snows that bury the rest of the region, making food plants inaccessible. Without access to Red Rim, the pronghorn would starve.

The rancher fenced off the land because he wanted it exclusively for his cattle. However, because Red Rim included Bureau of Land Management properties, which cannot be legally fenced for private purposes to the exclusion of wildlife, the U.S. District Court in Wyoming ordered the rancher to take down the fence. No permanent damage was done to the pronghorn, though television viewers who saw news clips of the snow-whipped animals dying beside the fence were treated to a look at the sort of tragedies that arise when wildlife needs clash with human ambitions.

Another problem for pronghorn is that coal underlies much of its range. Ranches in Wyoming, which has the largest single pronghorn population of any state, are already losing ground to strip mining. As demand for coal increases, Wyoming pronghorn will lose more habitat to strip mining. Mining will also bring more people into pronghorn territory, augmenting the problem of roadkills and poaching.

Conversion of grasslands to crops is causing rapid loss of pronghorn habitat and, if not stemmed, may become the pronghorn's greatest problem. Moreover, changes in the beef market could result in a jump in the number of cattle grazing on pronghorn feeding grounds. Aside from further habitat degradation by livestock, increased numbers of cattle can lead to increased numbers of fences. Fences pose a threat because pronghorn crawl under them rather than jump over, so their movements can easily be blocked with low strands of wire. Wildlife biologists recommend that ranchers build fences designed to permit the passage of pronghorn, but rancher cooperation cannot be counted on if a beef boom occurs.

Under current conditions, the pronghorn is secure against declines. But how long current conditions will prevail is anyone's guess. Humanity persuades itself that it has done well if it can increase a species' numbers for a decade or two, but nature charts its course in millennia.

CHAPTER 4

Waterfowl

They came with the waning of each year from the distant nesting grounds of northern Canada, trailing in sinuous formations across the skies of an entire continent. They followed routes mapped by their instincts, moving ever southward. Along the way they stopped to feed and rest in countless prairie marshes and woodland swamps and in the slow waters along river banks. Great clouds of both ducks and geese billowed over the vast chain of marshes and lakes that sprawled the length of California's Central Valley. Canvasback ducks darkened the skies when they flew over Chesapeake Bay, the pounding of their wings rumbling like distant thunder. When they settled on the bay they whitened its surface like new-fallen snow.

Ducks, geese, and other marsh creatures swarmed the far-flung wetlands of eastern Iowa and western Illinois. From Momence, Illinois, nearly to South Bend, Indiana, the 1,500 square miles of marshes along the winding Kankakee River clattered endlessly with the mutterings of ducks and geese. Geese congregated in the tens of thousands on Minnesota's Thief and Mud lakes, feeding on wild rice and wild celery. Countless ducks, as well as thousands of other waterbirds, summered in Montezuma Marsh, at the foot of New York State's Cayuga Lake.

In all, before the European settlers took over development of the continent from the forces of nature, waterfowl in the lower 48 states had more than 200 million acres of wetlands in which to court, feed, and nest, an area more than four times the size of New York State. These marshes, prairie potholes, swamps, bogs, and coastal salt marshes produced not only waterfowl, but some five thousand plant species, 190 species of amphibians, and a third of all the bird species in the United States. Nevertheless, the settlers thought the wetlands were worthless. In the middle of the nineteenth century Congress gave the states nearly 65 million acres of wetlands for agricultural development, and the nation began a spree of wetlands drainage that became a major factor in an abrupt decline in waterfowl numbers. By the early years of this century, California's Central Valley had been almost entirely drained, the bends of the Kankakee straightened by canals and the marshes destroyed, Thief and Mud lakes and the Montezuma Marsh lost to drainage, and eastern Iowa turned into farmlands.

In many cases, waterfowl were wiped out by poorly conceived efforts that soon turned to failure. The peat-bog bottom of Mud Lake was useless for growing crops. The hard-clay bottom of Thief Lake proved too flat to carry off heavy rains, so crops planted there flooded repeatedly. By 1929 the Minnesota legislature was contemplating a plan to turn the lake, drained only six years before, into a game refuge. Drainage of North Carolina's Mattamuskeet Lake, one of the Eastern Seaboard's most famous duck lakes, was initiated in 1914 and abandoned only ten years later because of high costs and continual flooding.

Wetlands devastation was not the only factor pressing waterfowl into decline. Egg collecting and sport and market hunting also contributed. Few states imposed limits on the number of ducks and geese that hunters could take. Those that did rarely enforced the laws. Individual sport hunters shot scores of ducks and geese in a single morning. A hunter might kill a thousand ducks in one winter. More interested in poundage of flesh than in prowess of gun, market hunters sought to increase their take by mounting large-bore shotguns, called swivel or punt guns, on the gunwales of boats. These shotguns were actually small cannons that could easily fire two pounds of lead pellets and kill as many as fifty birds with one shot. Since the birds were most easily killed at night, when they floated in dense flocks upon open water and could not see well enough for escape, market hunters would row quietly in among them after dark and shoot all night. In the morning, hawks would finish off wounded birds, and the sport hunting of daylight hours would begin. It was not uncommon in the 1870s for fifteen thousand waterfowl to be killed on Chesapeake Bay in a single day.

Such intensive hunting was practiced wherever waterfowl gathered to feed or nest. Ducks were sold by the market hunters for as little as a nickel apiece, though more commonly for a quarter. They ended up in the meat markets of large cities, hanging beside the carcasses of bison, deer, and elk or packed in barrels. Late in the nineteenth century the market hunters had yet to wipe out the passenger pigeon, but they had already laid to rest one species of waterfowl, the Labrador duck. Popular in eastern meat markets in the 1860s, the Labrador duck was last seen in 1875, when one was shot on Long Island, New York.

The double onslaught of wetlands drainage and overhunting sent waterfowl numbers plummeting. Canvasback ducks, for example, were rare on Chesapeake Bay by 1890. The decline of the canvasback did not lead to an abatement in hunting. Instead, it drove up the price of a canvasback to $6 or $7. By the opening of the twentieth century, waterfowl numbers were so low nationwide that drastic measures seemed necessary if the birds were to be saved.

THE FEDERAL SOLUTION

The most drastic measure was federal protection. At the time, the federal government had a scant role in wildlife conservation. Mainly, through the Department of Agriculture, it provided farmers with information on how to kill birds that damaged crops. The actual management and protection of wildlife was considered the legal domain of the states. This idea was so thoroughly entrenched in the U.S. legal system that virtually any federal law passed for wildlife protection would have been ruled unconstitutional in the courts.

Nevertheless, it seemed unlikely that the states would provide waterfowl, or any other animals sold in the meat markets, with adequate protection. As late as 1901, nine states or territories still permitted year-round waterfowl hunting. Only twenty-seven states had banned the use of punt guns, and only sixteen had prohibited the use of boats in waterfowl hunting. About half the states or territories still permitted the hunting of waterfowl at night. Only seven required resident hunters to carry hunting licenses.

The few wildlife protection laws that did exist in the state books were poorly enforced. Game rangers were few in number and could not effectively police the vast areas assigned to them. Moreover, many early rangers were themselves poachers or were tolerant of local poachers, who were often old friends.

It became increasingly clear to conservationists that the only way waterfowl protection could be reasonably accomplished was through federal regulation. They began searching for a way around the legal obstacle of states' rights, touching off a series of legislative battles that would change immeasurably the federal role in U.S. and even international conservation. Those battles in Washington, D.C., show how fights in wildlife conservation often are won not in the field but in legislative chambers.

In the forefront of the controversy were Massachusetts senator George Hoar, who wanted to end the slaughter of songbirds for the meat markets, and Iowa representative John Lacey, whose interest lay more in protecting game birds. They believed that the federal government's right to regulate interstate commerce gave it the authority to put controls on market hunting. It took them three years to shepherd a wildlife protection law through Congress, but they finally succeeded on May 25, 1900, when President William McKinley signed the Lacey Act into law.

The Lacey Act was the federal government's first major foray into the realm of wildlife protection. It banned interstate transport of wild animals killed in violation of state law and permitted the states to prohibit the import of game killed legally in other states. It also authorized the secretary of Agriculture to do whatever was necessary, provided that the states agreed, to restore wild bird populations.

It soon became apparent that the states were not interested in cooperating with the federal government in waterfowl management. It was also obvious that protection of migratory birds such as waterfowl, which range across state and national boundaries, would never be achieved consistently unless placed in federal hands. Consequently, in 1904 Pennsylvania representative George Shiras III introduced a bill that would have placed migratory game birds under federal control. President Theodore Roosevelt supported the bill, but that was not enough to save it from a quick death in Congress on the grounds that it was unconstitutional. Shiras and others continued to work for a federal migratory bird law, and one finally was passed in 1913. Called the Weeks–McLean Migratory Bird Act, it was soon the subject of two federal district court cases in which it was ruled unconstitutional. The judges argued that game was the property of the states and could not be federally regulated. One of the cases made it to the Supreme Court. By that time, the Department of Agriculture believed the federal case would be lost. However, once again a route around legal obstacles was discovered.

It had long been recognized that the federal government has the right to enter into treaties with other nations and that federal treaties are supreme over state laws. So officials in the Department of Agriculture urged officials in the State Department to sign a treaty with Great Britain, which would be acting on behalf of Canada, for the protection of birds that migrate between the United States and Canada. The treaty was signed on August 16, 1916. About two years later Congress passed the Migratory Bird Treaty Act, which gave federal agencies the authority to work for the goals outlined in the treaty. At that point, the Supreme Court dropped the case involving the Weeks–McLean act.

The treaty act gave the federal goverment the right to put closed seasons on waterfowl and other migratory bird hunting and to enforce laws for migratory bird protection. Passage of the act had immediate effects on waterfowl hunting as the federal government quickly reduced the length of open seasons. For example, in the opening decade of the century waterfowl hunting was open 225 days per year in Illinois. Federal regulations cut this to 105 days. Night hunting was banned nationwide.

Within two years after the act was passed, the federal government was again in court defending its right to protect migratory birds. The case had been started by the state of Missouri, which was trying to stop a federal game warden named Ray Holland from enforcing federal bird laws in the state. This time the Supreme Court came out strongly in favor of federal wildlife protection. It pointed out that since migratory birds travel freely across the states, the states are not in possession of the birds. And possession, the court declared, is the first step in ownership. The states therefore could not be said to own the game. The court also said that while the states may regulate the killing and sale of migratory birds, their authority to do so did not supersede federal treaty-making powers. The court also pointed out that were it not for the federal treaty "there might soon be no birds for any powers to deal with." The case of *Missouri v. Holland* proved crucial to future federal wildlife management because it broke the states' exclusive control of wildlife and indicated that the federal government might pursue wildlife protection even more aggressively. Since those early years similar bird-protection treaties have been signed with Japan, Mexico, and the Soviet Union.

The advances being made in waterfowl protection in the legislative arena were matched by advances being made in the field. Little was known about waterfowl early in the century. Indeed, little could be known, given the vast distances over which the birds traveled and the technological limitations that plagued wildlife managers at the time. But progress took place rapidly.

One of the first tools to be developed was banding. This involves capturing wild birds, putting small metal bands around their legs, and releasing them. The bands bear numbers that can be used to determine where and when a bird was banded, should it ever be captured again. Banding began no later than the early 1800s, when wildlife artist John James Audubon used silver wires in his own banding study. As a scientific tool, banding gained increasing importance in the opening years of the twentieth century. Between 1914 and 1916, one early wildlife manager banded 1,200 ducks on the marshes of Great Salt Lake. By 1922 another early manager, Fredrick Lincoln, was banding two thousand within a single month. By 1929, nearly half a million birds of all species had been banded in the United States and Canada.

Lincoln carefully examined data that came in from banded waterfowl that had been recovered. The data provided vital information. By 1930 it showed conclusively that waterfowl did not fly randomly over the continent. Instead, the birds migrated across the United States and Canada by traveling distinct corridors. Moreover, the waterfowl within each corridor tended to migrate between roughly the same wintering grounds in the south and the same nesting areas in the north.

Lincoln called the corridors flyways. He established that birds in each of the four flyways were more or less isolated from those in other flyways. Although some exchange might occur between flyways, they basically represented separate waterfowl populations. This meant that ducks, geese, and swans throughout the nation could no longer be treated as one large population. Managers would instead have to look at each flyway individually in making decisions about how waterfowl should be managed and protected.

The 1920s was a fertile time for waterfowl research, and this research led to an alarming discovery. It was made by Aldo Leopold, the man who made wildlife management history with his work on New Mexican and Wisconsin deer. In 1928 Leopold surveyed the waterfowl of eight north-central states, covering his costs with funds from the firearms industry. He found that despite federal protection, waterfowl numbers again were shrinking. He placed the blame on poorly controlled hunting, lack of surveys and of plans for waterfowl conservation programs, problems in cooperation between state and federal agents in enforcing migratory bird laws, and loss of wetlands through both drainage and the drought that was desiccating important prairie nesting areas.

As waterfowl declined into the 1930s, more restrictive controls were forced upon hunters reluctant to see their bag limits and season lengths tightened. Use of live decoys to attract ducks was banned, as was use of feed as bait. In 1935, federal regulations prohibited the killing of migratory birds with shotguns that held more than three shells.

Although losses of birds to hunters could be controlled to some extent, losses caused by the disappearance of wetlands were more difficult to remedy. Consequently, it was wetlands loss that most troubled Leopold, and he quickly pushed for the protection of waterfowl habitat. In a report he helped prepare in 1930 as chairman of the American Game Policy Committee, he outlined what he saw as the solutions to waterfowl problems: a nation-wide system of refuges where the birds would find respite from hunting; better federal, state, and international cooperation; and more research. The report, a reflection of Leopold's ideas, also lamented the lack of trained wildlife management person-nel. It pointed out that most managers were self-trained and lacked both scientific knowledge and management skills. It sug-gested that state wildlife agencies and universities, with guidance from the federal government, should undertake the financing and training of future wildlife managers.

Four years after the report was published, Leopold stumbled upon the chance to put his recommendations into action. Presi-dent Franklin Roosevelt appointed him to a committee that was supposed to create a program for wildlife restoration. The impe-tus for appointing the committee was the drought that was then destroying wetlands throughout the nation, sending waterfowl into decline. Called the "duck committee," the group was sup-posed to make recommendations that would permit the recovery of waterfowl and other wildlife. Its other members were Tom Beck, managing editor of *Collier's* and an early waterfowl advo-cate, and editorial cartoonist and conservationist Jay N. "Ding" Darling. Beck's goal was to abolish the Bureau of Biological Survey, the federal government's wildlife management agency, on the grounds that its personnel were untrained and incompe-tent. Leopold and Darling believed the agency simply needed better leadership. Leopold fought Beck on this point, Darling mediated, and the committee eventually was able to come up with a list of suggestions that included Leopold's earlier recommendations for a refuge system and a federal aid program to enhance the professionalism of wildlife managers.

Leadership at the Bureau of Biological Survey improved when the duck committee disbanded and Darling was named the survey's new chief. He quickly pushed for the goals outlined in the duck committee recommendations by ushering through Congress a law called the Migratory Bird Hunting Stamp Act, usually shortened to the Duck Stamp Act. This law required all waterfowl hunters over sixteen to purchase a federal waterfowl-hunting stamp in order to hunt. The money raised was earmarked for the establishment of national wildlife refuges. Thus Darling, working on ideas expounded a few years earlier by Leopold, was key in establishing the National Wildlife Refuge System. Designed originally for waterfowl protection, the system today encompasses more than 440 refuges and nearly 90 million acres vital to many types of wildlife. In the 1930s, when drought kept waterfowl numbers low, refuges established for migratory bird protection were closed to duck and goose hunting. Today, most are open to hunting of all sorts. In 1956 the purpose of federal refuges was expanded when Congress authorized the secretary of the Interior to acquire refuge lands for the protection of all forms of wildlife. Although duck stamp funds provided the impetus for establishment of the National Wildlife Refuge System, about two-thirds of the refuge lands outside of Alaska were obtained by withdrawal from federal lands rather than by purchase of private lands with duck stamp money.

Darling also succeeded in 1935 in creating a new cooperative educational program in which state land-grant colleges and wildlife agencies would work with federal wildlife managers to develop research projects and train new managers. He persuaded several firearms companies to pay for the program until federal money became available. This program has become one of the most important elements in U.S. wildlife research. Cooperative units are now located at land-grant colleges throughout the nation, and a similar program has been set up for work with fish and fisheries.

Within five years after the American Game Policy report appeared, concern for waterfowl had been translated into programs of lasting benefit to a wide varity of wildlife species. Development of the refuge system, improvements in wildlife management education, and restrictive hunting seasons helped waterfowl survive the dry years of the 1930s. Also, during the Thirties new advances were made in waterfowl research when federal biologists started using airplanes to survey waterfowl populations over vast areas. By the end of World War II, airplanes had become an important tool to waterfowl biologists as they compiled increasingly reliable data on waterfowl numbers and distribution. Banding programs also steadily grew through the Thirties and into the Forties. The 400,000 birds of all species banded in the United States and Canada by 1929 exploded to 5 million by 1945.

In the late 1940s another crucial change came to waterfowl management. It had been established nearly twenty years before that North American ducks, geese, and swans are divided into four populations, each of which lives primarily within one of four flyways. Early managers did not know this, however, so at first, seasons were set nationwide. Later, three or four management zones were created. Each zone ran horizontally from the East Coast to the West Coast, with seasons for the northernmost zone starting and ending earlier than seasons in the southernmost zone. In contrast to the zone system, the flyways generally run vertically north and south. The east–west system made it difficult for special considerations to be given to the birds of any particular flyway. Managers began to realize that this was an inadequate approach to the birds' needs. For example, hunters on the Atlantic Coast were given the same bag limits and season lengths as Pacific Coast hunters, although the numbers of waterfowl and hunters in each area differed considerably. In 1947 the U.S. Fish and Wildlife Service adopted the flyway system as the basis for its management decisions. In the early 1950s the states, seeking a greater voice in waterfowl management, established with federal cooperation four flyway councils made up of federal and state officials to coordinate waterfowl research and management within the Atlantic, Mississippi, Central, and Pacific flyways. The flyway management system is still in use today. The flyways, though they reflect biological divisions, are primarily administrative. Some birds shift between flyways. Tundra swans, for example, migrate from Alaska to the East Coast, and waterfowl in the Atlantic Flyway are not biologically independent.

It was not until 1955 that a consistent program was established for monitoring waterfowl populations. This is the annual breeding-duck population survey conducted by the federal Fish and Wildlife Service in cooperation with the Canadian Wildlife Service. The survey attempts to establish a baseline for duck populations by estimating the populations of ten species that breed in Alaska, Canada, and the lower forty-eight states. Not all breeding areas are counted, so the survey does not tally the total breeding population of each species. But the survey does cover a substantial portion of the duck population. When figures for several years are compiled, they reveal trends in duck numbers. Additional information and population trends are also derived from winter surveys. Goose and swan populations are similarly monitored in annual fall and winter counts.

WATERFOWL TODAY
. . . AND TOMORROW?

For the past half century and more, waterfowl have been among the most intensively studied, protected, and monitored of all game animals. Concern for waterfowl was among the wildlife issues that spurred the growth of wildlife management in the United States and created a federal role in wildlife conservation. The National Wildlife Refuge System sprang up around efforts to protect waterfowl habitat. Since 1934, the sale of eighty-eight million duck stamps has raised about $285 million for the purchase of waterfowl habitat. In addition, Ducks Unlimited, established by waterfowl hunters in 1937, has raised $490 million for the protection of wetlands in Canada, the United States, and Mexico. But despite all this attention, duck populations nationwide are steadily declining. The federal breeding-duck survey tallied an all-time low of 30.8 million in 1985. It rose to 35 million in 1987—still 12 percent below the average population size for all years since the survey was started in 1955—but in 1988 fell to 16 percent below average because of the summer drought. This figure may be even lower than the population levels of the drought-ridden 1930s, when declining waterfowl numbers alarmed wildlife conservationists such as Leopold and Darling.

The decline has been especially precipitous for certain species. Mallards are 20 percent below the average of all breeding-bird population surveys between 1955 and 1988. The blue-winged teal is down 25 percent and the pintail 54 percent. Most figures were reduced severely by the summer drought. But even before the long hot summer of 1988, many species were slipping. In 1987 the black duck, once one of the more common species east of the Mississippi and the most popular bird in the hunter's bag, numbered 17 percent fewer birds than it had the year before. The black ducks of the Atlantic Flyway tumbled 14 percent, and the Mississippi Flyway population 23 percent. Overall, black ducks may number less than half of what they did in 1950.

The causes for the long-term duck decline are complex. It is most likely that a combination of factors are working together to diminish duck numbers. It is also likely that different species are being suppressed for different reasons.

Agricultural development certainly has been a major cause of the waterfowl decline. A study in North Dakota has shown that ducks have greater nesting success on lands that are not grazed by livestock or mown for hay. Nest densities doubled on ungrazed lands, compared with lands that were grazed. "Elimination of grazing and mowing activities will result in increased waterfowl production," the researchers concluded.

All relevant studies also have shown that duck production is smaller on lands that are tilled every year. Twenty-five percent of the waterfowl that nested on untilled lands nested successfully, one study found. On tilled lands, the success rate was only 17 percent. Untilled lands were preferred by 72 percent of nesting ducks, except for pintails, which favored tilled lands. Only 15 percent of mallards, gadwalls, green-winged teal, and blue-winged teal nested on tilled land. A great deal of nesting habitat in the Canadian and U.S. prairie pothole regions, where at least half of the land is tilled, has been lost to drainage to create more agriculture land.

Loss of wetlands throughout the continent is without doubt another major cause of duck declines. Wetlands drainage can scarcely be said to have abated since the turn of the century. The amount of wetlands lost varies from state to state but is always vast. California and Iowa over approximately the past hundred years have lost more than 95 percent of their wetlands. Minnesota, touted to the tourist trade as the Land of Ten Thousand Lakes, has lost more than 90 percent of its marshes and swamps. Eighty percent of Mississippi's hardwood-forested bottomlands along rivers have been lost, mainly to the development of dams and reservoirs, and the U.S. Army Corps of Engineers—in what has been called the largest drainage project in U.S. history—is planning to drain half of the state's remaining 4 million acres of hardwood bottomlands. North Dakota, in the heart of the prairie pothole region that produces 80 percent of the ducks bred in the United States, has lost half its wetlands. In the late 1970s, North Dakota's elected state officials attempted to crush Fish and Wildlife Service plans for acquiring wetlands in the state to protect them from drainage for agriculture. Only court action permitted the service to elude the attempts of state officials to block federal acquisition plans.

The United States has destroyed about half of its wetlands, and Canada has nearly matched that pace. Total wetlands loss in the provinces of Alberta, Saskatchewan, and Manitoba—whose prairie potholes are among the most productive waterfowl breeding areas on the continent—is estimated at 40 percent. More losses lie ahead as farmers continue to drain potholes and marshes to increase agricultural acreage.

In the United States, an estimated half-million acres of wetlands are being lost every year. About 90 percent of the wetlands lost from the mid–1950s through the mid–1970s were drained for agriculture. But other causes account for considerable amounts of lost wetlands. In Louisiana, oil companies have cut thousands of miles of canals into the freshwater delta marshes at the mouth of the Mississippi River. The canals, built for the barge traffic that is part of the process of putting in oil wells, have allowed sea water to inundate freshwater marshes that were among the nation's most crucial for waterfowl and other wildlife. Louisiana is losing about 40,000 acres of delta marsh yearly.

Wetlands reduction is critical to waterfowl because it limits nesting sites, forcing the birds to nest in marginal areas that do not give them adequate protection from predators such as raccoons, skunks, and foxes. Moreover, as predation increases, managers of waterfowl refuges attempt to kill off the predators, compounding the insult to the natural environment.

Wetlands shrinkage also forces migrating waterfowl to crowd together in the limited number of remaining rest stops. Highly concentrated flocks are more vulnerable to hunters, elevating the number of birds killed unless increasingly restrictive hunting regulations are adopted. Crowding increases the spread of diseases such as cholera, which apparently spread to wild birds from domestic fowl sometime in the 1940s. As a result, epidemics can wipe out thousands of birds at a time. In March and April of 1980, an outbreak of avian cholera in Nebraska's Rainwater Basin killed nearly a hundred thousand waterfowl.

Ironically, some efforts to augment wetlands have created new problems. The classic case comes from California's central valley, an area vitally important to the ducks, geese, and swans of the Pacific Flyway. Seventy percent or more of the flyway's pintails, snow geese, tundra swans, white-fronted geese, shovelers, and gadwalls winter there. So do all of the Ross's geese and the endangered Aleutian Canada geese. Something over a century ago, the valley had 4 million wetland acres. Today less than 300,000 remain. Needless to say, the Fish and Wildlife Service would like to increase that amount. Consequently, when in 1972 the federal Bureau of Reclamation offered the service the chance to manage for waterfowl some bureau ponds formed by water running off agricultural lands, the service jumped at the chance. Thus was born Kesterson National Wildlife Refuge. It soon turned out, though, that a toxic element that occurs naturally in soils around Kesterson was building up in the agricultural runoff and accumulating at the refuge. In 1983 this element, called selenium, started killing off refuge birds. Birth defects also became rampant. A 1984 study discovered that one of every five embryos in the nests of some species were deformed and that almost half the embryos were dead. The refuge subsequently was declared off-limits to waterfowl and other waterbirds. The refuge manager now attempts to scare off the birds by periodically firing shotguns into the air. Discovery of the Kesterson pollution problem stimulated investigations into pollution at other refuges. Refuges in Utah and Nevada are suspected of having similar problems, and refuges in California's northern central valley that use agricultural runoff are almost certain to face trouble.

A more extensive pollution problem may be building up from acid precipitation. A survey of independent studies released by the Izaak Walton League of America suggests that acidification of ponds by acid rain may be adversely affecting waterfowl in several ways. For instance, acidification may be killing off foods needed during the breeding season, such as calcium-rich snails. As foods diminish, competition between fish and ducklings for the same food sources intensifies. Fish, more mobile than ducklings, win.

The number of fish also can be reduced when the ponds in which they live are acidified by acid precipitation. This can be crucial to ducks that feed on fish. One study found that in Ontario, common mergansers, which are fish-eating ducks, produced five to eight times fewer young in areas with high acidity.

Studies in the United States have yielded similar results. The Fish and Wildlife Service has shown that the death rate among young black ducks on experimentally acidified wetlands was more than three times greater than among ducklings of the same age reared on nonacidic wetlands.

Aside from cutting into food supply, acidification is thought to affect ducks by increasing the amount of toxic metals in pond waters. This happens because higher acidity permits toxic metals such as aluminum, common in most soil, to dissolve and enter the water. Once in the water, the metals are absorbed by snails and fish. When waterfowl feed on these animals, the toxins accumulate in the birds' tissues. Eventually, ducks, geese, and other waterbirds can become weakened and, unable to fend for themselves, die.

Loss of wetlands to drainage and pollution—human-caused problems—probably rivals naturally occurring droughts as a major cause of waterfowl declines. The birds generally can recover from natural droughts, since arid conditions are not permanent. In most cases, however, wetlands drainage has been permanent, and no species can recover from permanent habitat loss.

Another, though perhaps less critical, factor affecting waterfowl numbers is hunting. U.S. hunters from 1980 to 1985 annually killed about 1.9 million geese and 9.6 million ducks. Hunting seasons and limits are usually developed under the premise that hunters kill only animals that would die anyway of some other cause, such as lack of food during the hard months of winter. However, some evidence suggests that this may not always be true, particularly for localized waterfowl breeding populations. A study of mallards that breed in Minnesota indicated that losses to hunting were at least in part additive to nonhunting causes of death. This means that, in this particular study, a portion of the Minnesota-bred mallards killed by hunters would otherwise have survived to breed more birds the following spring. Hunters, the study showed, killed about 60 percent of the mallards breeding in Minnesota, with other causes of death accounting for another 20 percent. The total mortality of 80 percent meant that the local breeding population could not sustain itself.

A 1987 federal study that assessed the impact of hunting on waterfowl concluded that in some years hunting nationwide kills up to 25 percent of the fall flight of mallards. The study asserted, "In view of the relatively low populations in recent years, mallard harvest rates may be cause for concern." The study pointed out another ominous sign. The number of ducks killed during hunting seasons from 1979 to 1984 was nearly the same as the number killed from 1968 to 1978, even though duck populations in the early 1980s had declined to the lowest levels on record. The federal government has responded by reducing bag limits as duck numbers have sunk since 1985.

Research on black ducks, published in the *Journal of Wildlife Management* in 1988, indicated that young male black ducks "exhibit additive mortality" from hunting. Data for female black ducks were inconclusive. The authors declared that thirty years of banding and recovery data still did not enable them to determine how hunting is affecting the black duck population. This is a chilling statement. If the affect of hunting upon black ducks is unknown, then black duck seasons are based on questionable information. The authors themselves suggested that in the absence of a conclusion, managers should perhaps be conservative in setting limits. Indeed, black duck bag limits have been reduced up to 40 percent since 1983.

A side effect of hunting is the lead poisoning of waterfowl. Ducks, geese, and swans are usually hunted with shotguns, which scatter lead pellets into the environment with every shot. The pellets settle into marsh waters and mud, where feeding birds find and eat them. The quantity of lead involved may not sound like much—perhaps an ounce and a half with each shot. But the amount must be multiplied by the number of shots taken. A 1952 study found that hunters on Horicon Marsh in southwest Wisconsin fired twenty-eight shots for every goose killed. That would suggest that each goose killed in the study represented nearly two pounds of lead scattered over waterfowl feeding grounds. Because hunters at the marsh killed more than eighty thousand geese from 1953 to 1961, they presumably put more than 80 tons of lead into the environment at this one site.

When birds eat lead shot, the pellets are ground in their gizzards. The crushing of the pellets makes it easier for the digestive system to erode the lead, permitting it to enter the bloodstream. Poisoned birds become weak. Paralysis may set in, but a weakened condition by itself can cause death from starvation.

The deleterious effects of lead have been known since at least the end of the nineteenth century. George Bird Grinnell wrote about it in *Forest and Stream* magazine in the 1890s. A search for a substitute for lead shot began in the 1930s, but nothing happened to improve the situation. Then, during the 1970s, mounting evidence indicated that lead poisoning was causing serious waterfowl losses, even cutting into breeding stock. Steel shot was proposed as an alternative, and tests of steel shot were undertaken. Under experimental conditions, shells loaded with lead shot killed more geese than steel shot and crippled fewer birds. However, when tested by hunters in the field, no significant differences were found in the killing and crippling rates of lead and steel.

In the late 1970s the Fish and Wildlife Service began to cut back on the use of lead shot by establishing "lead-free zones" in parts of twenty-four states. This attempt to increase the use of steel shot was soon blocked by hunters and by gun enthusiasts represented by such groups as the National Rifle Association. They claimed that steel shot had different ballistic qualities and that its use would require them to learn new shooting habits. They persuaded Congress to make it illegal for the Fish and Wildlife Service to spend money enforcing the lead-shot ban unless the state in which the regulation applied agreed to the rule. This stifled the lead-shot bans because most state wildlife managers succumbed to hunter pressure and opposed the lead-free zones. Wrote Frank C. Bellrose, the dean of U.S. waterfowl biologists, in a review of the lead-shot story, "Many state administrators, even if they favored the use of steel shot, were unable to withstand the pressure placed upon them by hunters opposed to steel shot. This pressure was especially evident when a neighboring state did not support the steel shot regulation."

Federal efforts to ban lead shot were revived in 1984, when the National Wildlife Federation brought a lawsuit against the Fish and Wildlife Service on the grounds that lead-shot ingestion was killing bald eagles, an endangered species that fed on wounded waterfowl. Lack of a ban, the suit said, was a violation of the federal Endangered Species Act. The courts upheld the case and lead-shot bans were initiated in 1985 in a number of areas where the potential for lead poisoning was particularly high. Further lead-shot bans were slated for ensuing years as part of a plan to eliminate lead-shot use within a few years.

PROGRESS AND PROTECTION

Clearly, waterfowl are facing some serious problems. It takes little imagination to conclude that the compound effect of these problems lies at the root of the waterfowl decline. Nevertheless, some progress is being made.

Four species of Pacific Flyway arctic-nesting geese—the black Brant, dusky Canada, cackling Canada, and Pacific white-fronted goose—were in decline for some years. Some evidence suggests that their populations have been diminished by over-hunting in the Pacific Northwest and on the birds' nesting grounds in Alaska, where in spring Native Alaskans hunt the birds and their eggs for food. Bag limits in the United States were reduced in an effort to halt the decline, but for several years little was done to keep Natives from taking nesting birds. The Fish and Wildlife Service was reluctant to prosecute in arctic Alaska because the Natives had threatened to kill massive numbers of the birds in retaliation and law enforcement in the vast nesting area was nearly impossible. However, since 1985 Alaska Natives have agreed to cut back on the spring take. The turnaround came after a Fish and Wildlife Service educational campaign showed the Natives that they would lose the birds forever if they did not restrict their take. Since 1986, biologists have seen encouraging improvement for cacklers, white-fronts, and brant, as well as for an endangered species, the Aleutian Canada goose, which is likely soon to be reclassifed to less critical threatened status. Duskies are low but stable.

Wetlands protection received a boost when the 1985 farm bill was passed by Congress. The bill included a swampbuster amendment that said that all federal farm-program benefits would be taken from any farmer who drained wetlands to create agricultural lands for commodity-crop production. The Fish and

Wildlife Service hopes the amendment will help protect some 5 million acres of wetlands in the lower forty-eight states, including prairie potholes and lower Mississippi valley bottomlands. There is little doubt, however, that in highly productive agricultural areas the swampbuster amendment alone will not be enough to stop unnecessary drainage.

Perhaps the greatest advance for waterfowl management is the North American Waterfowl Management Plan, in which the United States and Canada have agreed to cooperate more intensely on waterfowl management for the next fifteen years. The plan also calls for a large role for private organizations, such as Ducks Unlimited. It emphasizes primarily habitat protection, but also asks for the appointment of special committees to monitor waterfowl protection and research; seeks acquisition of waterfowl habitat, notably on the Canadian prairies and in California's Central Valley; and puts special emphasis on the protection and management of the black duck and arctic-nesting geese.

The plan is vast in scope. In addition to its call for intense cooperation among the Canadian and U.S. governments and the private citizens of both countries, it sets population goals for ducks and geese. It also calls for private organizations and public agencies to raise $1.5 billion over the next fifteen years to fund implementation of the plan's goals.

If the plan is pursued aggressively, it will prove to be a major step for waterfowl conservation in North America. But effective implementation will require that agriculturalists place waterfowl conservation before profits; that hunters put the interests of ducks, geese, and swans before their desire for liberal bag limits and long seasons; and that wildlife biologists in the United States and Canada put management needs before jurisdictional jealousies. The success or failure of the plan will be a measure of U.S. and Canadian commitment to wildlife conservation.

CHAPTER 5

On
Wildlife
Management

I worked for four years in the public information division of a state wildlife agency, a place administered primarily by people with degrees in wildlife management. Frequently among my assignments was the writing of a magazine article, television spot, radio-program script, or brochure that would praise the role of the hunter in conservation. Written from the perspective of a wildlife manager, this generally could take one or all of three approaches—pointing out that hunters were at the forefront of the conservation movement and in fact responsible for its creation; showing that because hunters pay for most of the conservation programs in the United States, they are a hell of a bunch of guys and should be given the biggest voice in setting wildlife management policy; and/or showing how, thanks to the hunter's money and influence, the world has been made safe for wildlife. The proof of this last argument was the many deer, waterfowl, turkeys, and other game that hunters kill each year.

No one can deny that hunters, though primarily hunters as wealthy eastern sportsmen, were at the forefront of conservation late in the nineteenth century. But to say that they were the exclusive progenitors of conservation is to overlook the efforts of nonhunters in the early years. These people included such notables as John Muir, the Sierra Club founder who saved wildlife through his relentless efforts to save wildlands. Nonhunters also fought successfully at the turn of the century for a bird protection law in order to save egrets and other beautifully plumed species from being shot into extinction to supply meat for the markets and feathers for the millinery trade.

The bond that links wildlife managers to the hunting community was forged long ago. At the turn of this century, when wildlife managers took on the tremendous challenge of recovering vanishing game species, they had few of today's sources of funds. The laws passed in the ensuing decades provided the funds by tapping the pockets of hunters, welding a close attachment between managers and hunters and establishing a sort of tradition in wildlife management.

Today's wildlife managers are locked into that bond and steeped in that tradition. They believe that because hunters are required by a variety of state and federal laws to pay hunting fees, they are their closest allies. This is why writers who work for state wildlife agencies are assigned to produce magazine articles, radio and television programs, brochures, and a host of other pieces that repeat endlessly that hunters are the nation's true conservationists.

Though the financial relationship between hunters and wildlife managers is compelled by law, no legislation has been passed requiring nonhunters to pay fees for wildlife conservation. Nevertheless, nonhunters pour millions of dollars into conservation each year by joining conservation groups that work for better wildlife protection. Annual membership fees for some groups hover around $30, more than a hunter would expect to pay for a resident hunting license and a deer permit. These groups, which include organizations such as the National Audubon Society, Sierra Club, Defenders of Wildlife, and the Humane Society of the United States, have been instrumental in the passage of such landmark wildlife protection laws as the Endangered Species Act, Marine Mammal Protection Act, Wilderness Act, and National Environmental Policy Act. Certainly, hunters join these organizations, too. Aldo Leopold was among the founders of the

Wilderness Society, which is devoted to saving the last of the nation's roadless areas. But to ignore the nonhunting membership of these groups is to ignore the profound contribution nonhunters have made to wildlife conservation.

The idea that hunters created and pay for conservation would be a harmless conceit were it not that wildlife managers, by accepting it, have permitted the hunting community to dominate and at times control wildlife-management policy. And this itself would be harmless were it not that the most vocal hunters, those most likely to foist their ideas upon wildlife managers, seem always to be the segment of the hunting community with the least interest in true wildlife conservation. Their primary interest lies in the preservation of bag limits and in opening more areas to hunting, often regardless of the biological needs of the species they hunt. That this interest is too narrow to yield sound wildlife conservation, and stifles wildlife management, can be seen in the problems Leopold had with deer management in Wisconsin. It can be seen in hunter opposition eighty years ago to tighter limits on deer and waterfowl. It can be seen in the records of the 1950s, when hunters in some states successfully demanded the release of domestic and semidomestic turkeys even though professional biologists knew the releases were a waste of funds and a threat to the survival of the few remaining wild birds. It can be seen today in hunter opposition to steel shot and antlerless deer hunting in areas that need it.

A game division chief I talked to in 1980 said that the only solution to the duck decline would be to close hunting of the birds for five years. But, he said, hunters would never allow it. In response to hunter pressure, some southern states have opened special seasons during which deer may be pursued with dogs, even though many wildlife biologists are convinced that this controversial form of hunting is harmful to deer. In Florida, a portion of Loxahatchee National Wildlife Refuge had been closed to virtually all human intrusion for nearly thirty years, making it the northernmost portion of pristine Everglades. It was opened to hunting by the secretary of the Interior at the behest of the Florida Wildlife Federation, a hunter-dominated state chapter of the National Wildlife Federation, even though the refuge manager said the deer population was in balance with its environment and would not benefit from being hunted. Thus wildlife managers, by acquiescing to the idea that they must serve hunters because hunters pay for conservation, in many cases have rendered themselves powerless to take measures they know are needed for sound wildlife management.

The hunting community's stranglehold on wildlife management exists only because state and federal bureaucrats have helped create and perpetuate it. They have been content to remain bound by tradition, focusing management on game species and functioning as if hunters were the only possible source of funding. This miscalculation has kept wildlife managers from forging an alliance with nonhunting wildlife enthusiasts and from discovering avenues by which nonhunters could provide additional funds for wildlife conservation. Funds from nonhunters could become particularly important in the future, because annual hunting license sales figures indicate that the number of hunters in the nation is in steady decline. Also, when hunted species, such as waterfowl, decline, hunters tend to abandon their sport, which means that they stop paying the fees on which wildlife managers depend. Nonhunters presumably would provide a more reliable financial base.

Bringing nonhunters more intimately into the conservation decision-making process might bring wildlife and wildlife management more benefits than just money. Nonhunters presumably would be less likely than hunters to oppose such measures as tighter waterfowl bag limits and lead-shot bans, since their concern would be with more ducks in the sky, not more in the bag. They could prove a valuable ally when wildlife managers are seeking tighter protection for other game species as well.

Moreover, nonhunters presumably would seek to expand the work that wildlife biologists do to include more than game species. For although wildlife managers have accomplished what they set out to do nearly a century ago—recover game animals—they have been reluctant to go beyond that. This is because wildlife management is dominated by what conservationists call the "good ol' boy network," a coterie of people with a lifelong interest in hunting who have turned their hobby into their profession. They are a distinctive type of biologist, more like those who grow crops or raise livestock than like scientists. They even call the killing of game a harvest and a population of animals a crop. Unless populations are dwindling, their interest in wildlife is restricted primarily to knowing approximately how many animals are in the field and how many can be killed by hunters. They also have a major interest in knowing how to increase numbers of game animals, even if the measures taken to do so jeopardize the survival of nongame species. For example, they will burn or cut a portion of forest to create openings for deer, ignoring the needs of species dependent on untouched forest lands. Until now, managers have been able to follow their personal whims, placing limits on which species receive the benefit of their expertise rather than broadening the responsibilities of their field. These broader responsibilities would include more active protection and conservation of nongame species, animals that are of increasing concern to most members of modern society, whether hunters or not. Assuming a larger role in the protection of nongame animals would make wildlife management truly *wildlife* management, and not simply game management.

Nevertheless, wildlife management is widening its horizons, if only because expansion is being foisted upon it willy-nilly. Expansion began in 1966 with passage of the first federal Endangered Species Act. The law authorized the federal government to compile a list of species that seemed in danger of extinction. However, it provided no protection for the listed species, merely noting their precarious existence. In 1973 Congress enacted a new Endangered Species Act that protected listed plants and animals. Species may be nominated for listing by any interested party. The U.S. Fish and Wildlife Service is charged with determining which candidates should be listed, except that marine species such as whales are administered by the National Marine Fisheries Service. A candidate may be listed as endangered if in danger of extinction throughout all or a significant part of its range. A candidate likely to become endangered within the foreseeable future is listed as threatened. Protections for threatened species are not as tough as those for endangered.

The Endangered Species Act has been touted as the most powerful wildlife conservation law in the world. It made it illegal for U.S. citizens to harm, capture, or market animals that appear on the Federal Endangered and Threatened Species List. It helped provide funds to the states for work on endangered animals and plants and made it illegal for the federal government to fund any projects that harm listed species. The act was the first sign that conservation in the United States had advanced beyond the point of protecting only game animals.

Part Two

Endangered Species

CHAPTER 6

The
Gray
Wolf

Something about wolves—their cool yellow eyes, their nocturnal and predatory ways, their living in packs—makes writers wax mystical about them, even sensible writers who with any other species would leave all the esoteric similes at home in their type-writers. Give them a wolf to write about, and they often seem compelled to spiritualize and mysticize the animal into abstrac-tion, as if the creature by itself could not mesmerize a reader, could not attract normal human interest without the writer's invention. Such an approach may be a disservice to the animal, but it shows how thoroughly the wolf has entranced those who favor it.

Conservationists are not immune to the wolf's charms, either. Most reliable conservationists have at one time or another thought of the wolf as the ultimate wilderness symbol and be-lieved that its howl in the night is a sign that the listener has reached true wildlands. Where the wolf lives, according to this idea, wilderness reigns. This too is a measure of the sublime effect that the wolf has on its adherents.

It is relatively easy to succumb to these fancies, to go off to the Canadian north woods in quest of a mystical wolf. With expecta-tions such as these, a traveler can go to Ontario's Algonquin

Provincial Park on a clear, still night and attend a wolf howl. The park holds these affairs on Thursdays in August. The expectant traveler might presume the howl to be an intimate experience in the pine forest, with perhaps a handful of wolf enthusiasts gathering at dusk to cruise the back roads in three or four cars, stopping occasionally so that a park ranger can howl into the night, moving on to another site if a response does not come. It might take many tries in many places, the night growing chill around.

With literary expectations in mind, a traveler might indeed anticipate something like this. But he would be in for a surprise.

August 13, 1987. Dusk in an outdoor theater surrounded by aspens and jagged pines. People gather on low wooden benches that face a stage. The scent of cigarette smoke sharpens the air. Algonquin Park officials give a short lecture on wolves. Then they suggest that all go to their cars and prepare to hear wolves howl.

Some 1,480 people then repair to 370 cars. Gentlemen, start your engines. Park rangers direct the traffic out of the parking lot and onto Highway 60, the main thoroughfare that cuts across the southern section of the park. Highway 60 has been closed temporarily to through traffic so the wolf-howl contingent can get to the wolf-howl site, which they do with flawless precision under the equally flawless guidance of the park rangers.

It is dark. Stars spangle the sky. The 370 cars now line both sides of a narrow dirt road. Following instructions given at the outdoor theater, all participants stand beside their cars. No one speaks above a whisper. No car doors are slammed. As the lecturer at the theater had said, you will be surprised at how quiet 1,480 people can be when they do as told.

When everyone is in place, a ranger at the end of the long line of cars signals another ranger at the front of the long line of cars. The howl begins.

Two men make the howls. It is an ear-shattering sound to anyone standing near them, operatic, glass-breaking. The howlers have a distinct plan. They will howl several times at regular intervals. If the wolves do not respond, they will wait a few minutes, then howl several more times at regular intervals. After that, if the wolves do not reply, they will call it a night. The odds favor those who want to hear wolves, since park officials, leaving little to chance, have located the animals in the vicinity a night or two before. Also, it is a still night, which is essential since the cries of wolves do not carry far.

The wolves respond almost immediately from somewhere that seems far off under the night sky. The howls are wafer thin, delicate. They seem to shimmer, if sounds can be said to shimmer. The howls of the adults are mellow, those of the pups quite shrill, like yelps.

The sound ends after a few seconds. In the darkness the celebrants wait. Fifteen minutes go by and the rangers howl again. The wolves respond, the howls fade, the night is over. Under the rangers' careful instruction once again, visitors are guided on their way.

A pure wilderness experience? There are whole towns in the Midwest that do not have 370 cars in them, let alone 1,480 people. But it is all a matter of perspective, perhaps. To the wolves, for example, it may be very much a wilderness experience.

No matter. It is definitely a wolf–human experience, and as wolf–human experiences go, this one is certainly something new. For perhaps the first time, people are pursuing wolves out of interest, concern, even affection. That is a far cry from the traditional human approach to wolves, an approach that has purged the animal from almost all but the most distant reaches of North America.

DECLINE OF THE WOLF

At the start of the seventeenth century perhaps two million wolves roved across the continent. They came quickly into conflict with European settlers. In 1609, nine vessels from England docked at Jamestown, Virginia, laden not only with five hundred settlers but also horses, cattle, pigs, and sheep. Wolves ate the sheep. In 1630, Massachusetts established the first bounty on wolves, paying a reward for dead ones. In 1654 and 1655 Delaware required all resident Native Americans to turn in two dead wolves yearly as a sort of tax. At the start of the eighteenth century, wolf hunting was a pastime of young Virginia aristocrats, who often could be seen riding hard into town with live wolves dragged behind their horses.

Nearly a hundred years later, newly elected president George Washington complained in letters to the president of the Agricultural Society of Great Britain that wolves made the raising of sheep on the frontiers impossible, but in fact wolves were then already in decline. They had been extirpated from most of New England and within the next few decades would disappear from the East, hanging on longest in the more rugged reaches of the Adirondacks and the Alleghenies.

In the nineteenth century livestock were moved onto the western plains, increasingly so in the second half of the century. Between 1866 and 1883 nearly five million head of cattle were driven out of Texas and into the northern prairies. During that same period, hunters decimated the vast herds of plains bison. Killing off the bison was virtually a federal policy, designed to clear the grasslands for cattle and destroy the Native Americans dependent on the buffalo. Some states tried to protect the bison with closed seasons, but too much money was to be made selling hides and meat and too much land to be gained in their extermination.

With the bison gone, along with the elk, mountain sheep, pronghorn, and deer, the plains wolves, numbering perhaps three-quarters of a million at the start of the nineteenth century, turned increasingly to domestic cattle and sheep for the bulk of their diet. Ranchers fought back with guns, dogs, traps, and poisons. Former buffalo hunters became professional wolf hunters, called wolfers. Shortly after the turn of the twentieth century, the federal government joined the fray. At first the government acted in an advisory capacity. Then the Forest Service started killing wolves in the national forests. By the 1930s, federal wolf-control programs had been extended to private lands. The programs were successful. Today, outside of Alaska, wolves have only a scant toehold in the United States. Perhaps

1,200 survive in remote northern Minnesota, a forested area for the most part poorly suited to agriculture. Perhaps 50 remain in similar habitat in Michigan and Wisconsin. Perhaps another dozen or two, transients from Canada and their offspring, can be found in northwestern Montana. And that's it.

The gray wolf was added to the federal endangered species list in 1967. This earliest version of the Endangered Species Act did not protect listed animals. Listing was merely a statement that the species' existence was in jeopardy. Had that law provided legal protections, wolves might still exist today in the Southwest. For in the 1960s wolves were still occasionally reported in New Mexico and Arizona, though probably all were stragglers from Mexico. Eventually, they were hunted and killed to protect livestock. What happened to them fairly well encapsulates the fate of wolves throughout the West in the twentieth century.

The most intensive phase of wolf control in the Southwest, as elsewhere, began in 1915, the year Congress first apportioned funds to the Bureau of Biological Survey specifically for the killing of wolves. The conventional wisdom of the time suggested that three hundred wolves could then be found in New Mexico. When $20,000 was granted for the control of wolves in New Mexico and Arizona, it was promptly invested in two professional wolf hunters. This was the beginning of a program that would not end until the last wolf was killed in the Southwest in the 1970s.

During their first year of operation, the federal agents killed 69 wolves. The number tended to climb over the next four years to an all-time high in 1920 of 110. From then on, with some fluctuations, the numbers began to decline. In 1925 the agents killed fewer than 30. In his annual report for that fiscal year, J. Stokley Ligon, the man in charge of southwestern wolf control, wrote, "The passing of the wolf in New Mexico, as well as in other western states, is every year becoming more apparent. . . . The 'Lobo's' final exit from New Mexico has long been heralded. His stay, which has been far too long, seems fast drawing to a close."

Ligon was premature in his hope for extinction of the wolf in the Southwest, but certainly by the 1930s the wolves being killed in New Mexico were mostly animals that had crossed the border from Mexico, where a population still existed. They were few in number. The average annual kill in the 1930s was only fifteen wolves. In the 1920s it had been seventy-eight. From the 1930s on, the control agent's job was mainly one of patrolling the Mexican border and killing any wolves that crossed. The job came closer to completion after World War II, when a new poison, sodium monofluoroacetate, also called Compound 1080, was added to the wolfer's arsenal. Tasteless and odorless, the poison was especially effective against canines. By the end of the Forties, southwestern wolf numbers had been severely reduced. Consequently, from 1950 to 1953, the agents were able to kill only eight wolves altogether in Arizona and New Mexico. All were Mexican wolves. The only other wolves known to be killed in the Southwest in the 1950s were a pair poisoned with cyanide in Arizona.

The federal agents killed their last Southwest wolf in 1960. In 1961, a wolf was trapped north of Wilcox, Arizona, by a private citizen, and a bounty of $75 was paid on it by the Arizona Livestock Sanitary Board. No more were killed in Arizona for more than ten years. In New Mexico, the last known wolf was poisoned in 1970. Two killed in southwest Texas that same year— one was shot and one found dead in a trap—almost closed the record on the wolf in the Southwest, which meant that the entire western United States was empty of wolves.

It was three years after this that Congress enacted a new version of the Endangered Species Act that protected from destruction the species listed under it. The Mexican wolf, though extinct in the United States, was put on the list. This was done to protect the tiny handful that still survived in Mexico's Sierra Madres, despite great reduction by human encroachment on their habitat and by government control programs. However, in 1976 it appeared briefly as if the act might help preserve the wolf in the southwestern United States, too. That year wolf tracks and animals apparently killed by a wolf were discovered in the Aravaipa Valley region of Arizona. However, the wolf was soon illegally killed by a private trapper who was reputedly hired for $500 by local ranchers. No wolves have been seen in the Southwest since.

However, gray wolves may some day appear again in the border states. About thirty Mexican gray wolves, all of that subspecies known to survive, are being held in captivity under the authority of the U.S. Fish and Wildlife Service. The service has designed a plan to release some of these wolves, and Texas, Arizona, and New Mexico have been asked to submit lists of possible release sites. Unfortunately for wolf enthusiasts, Texas has refused to cooperate, and rancher opposition is stifling progress in the other states.

MANAGEMENT TODAY

Wolf management today is primarily a matter of how many wolves will be killed in a given area and under what circumstances. This is not entirely bad, however. The unrestrained slaughter that characterized wolf control in the first half of this century is almost certainly a thing of the past in the lower forty-eight states. The control measures that continue often constitute a kind of protection for the wolf, as management in Minnesota shows.

Minnesota's wolves were federally listed as endangered under the 1973 Endangered Species Act, making it illegal to kill or even harass them. However, among a part of Minnesota's human population there exists a long tradition of wolf hating. These people blame the wolf for killing livestock or deer. When the wolf was given complete protection, some of them took it upon themselves to kill and bury wolves on the sly. Sometimes, however, they were blatant about it, delivering carcasses of illegally killed wolves to the doorsteps of newspaper editors and legislators as a form of protest against the Endangered Species Act.

To help placate those vehemently opposed to wolf protection, the status of the Minnesota wolf under the Endangered Species Act was reduced in 1978 to threatened. As a result, problem animals, such as those that prey on livestock, could be killed by federal agents. This helped relieve concerns that total wolf protection would lead to widespread predation on livestock. Under this system, when a farmer reports the loss of livestock to wolves, a federal agent examines the livestock carcass to determine the cause of death. If the animal was in fact killed by a wolf, traps are set. This is done because wolves that kill livestock once will probably do so again. Wolves caught may be killed or taken into captivity for research.

Not all wolves in Minnesota's 30,000 square miles of wolf habitat are subject to trapping. The state's wolves are managed under a system of five wolf management zones. Zone 1 is a 4,400-square-mile area wedged between the Canadian border and Lake Superior. In this core recovery area, where little livestock is produced, wolf protection is the primary goal. In Zones 2 and 3, which cover some 6,000 square miles of public and wild lands, wolves that prey on livestock are trapped, although the area is considered part of the wolf sanctuary. Zones 4 and 5 comprise the rest of the state, and human land uses take priority there over wolf protection.

Where wolves are trapped, strict regulations govern control operations. Efforts have been made to weaken the regulations. For example, in 1983 the Department of the Interior tried to turn management of the wolf over to the Minnesota Department of Natural Resources, which planned to allow a sport trapping season on wolves and sale of the wolves' hides. Several conservation groups sued to stop this action, arguing that a threatened species cannot be hunted legally except in cases where local populations have become excessive. The federal judges who heard this case and the appeal agreed with the conservationists, and wolf management was left in the hands of the Department of the Interior. In fact, when conservationists pointed out that 250 to 450 wolves are killed illegally in Minnesota each year, one judge suggested that Interior should be trying not to open a hunting season on wolves but to protect wolves better.

Regulations for wolf control require that traps be set no more than a half mile from the farm where the livestock was lost and that trapping be stopped no more than twelve days after the last predating wolf is killed. Any pups that are trapped before August 1 must be released, for pups so young are not by themselves capable of killing stock animals.

Although wolves that kill cattle, sheep, horses, and pigs are removed as quickly as possible from the population, wolves in fact do little damage to farm animals. One study showed that of the twelve thousand farms in northern Minnesota, only about forty report livestock losses to wolves in an average year. In some cases the problem stemmed from poor handling of livestock. For example, many farmers left cattle overnight in unprotected pastures frequented by wolves.

The total average annual loss of livestock to wolves in Minnesota is about five head of cattle and a dozen sheep for every ten thousand of each animal grazed there. Farmers are compensated with state funds for these losses. From 1978 to 1980, twenty-five to thirty-one claims were paid yearly at an annual average cost of $21,000.

The wolf-control program and compensation for livestock losses have doubtless kept some farmers from becoming ardent wolf haters. But farmers who oppose wolf protection have found strong allies in a portion of Minnesota's sport-hunting community. Because their hunting fees help pay for wildlife management, many hunters have a proprietary attitude about wildlife. They see wolves as competitors for the white-tailed deer that is one of the most popular game animals in the state, and many of them fear that protection of wolves will lead to extermination of deer. Hunters even point to statistics showing that some deer populations have declined since wolf protection began. However, figures from recent deer-hunting seasons indicate that deer numbers are high throughout Minnesota.

Research shows that although wolves usually cannot reduce healthy prey populations, they can slow or stop the growth of populations already depressed or reduced by other factors, such as severe weather or lack of food. This observation has been at the heart of a controversial wolf-management system far removed from Minnesota's: the Alaska aerial wolf hunt.

Alaska is the only state that still has a large wolf population, some five thousand to six thousand animals. That is only half the number thought to live there five years ago, though it is not clear whether the revised lower figure represents a better count or a real decline. Regardless, the reduced population estimate has not daunted the Alaska Department of Fish and Game in its desire to cut down the state's number of wolves. The game department has been trying to eliminate wolves off and on over the past decade and a half. The reason is human competition for the caribou and moose that are the primary components of the Alaska wolf's diet.

Official wolf control in Alaska dates to 1915, the year the federal government first appropriated funds for killing predators. In 1948, the federal agents acquired a Super Cub airplane and a year later started intensive use of the plane in the aerial hunting of wolves. Aerial hunting involves flying over wolf habitat, spotting wolves below, and then flying low over the animals, which usually flee at the plane's approach. The pilot tries to come in close enough so that his partner can kill the wolf with a shotgun. Sometimes the pilot overreaches his flying ability or the gunman shoots out a wing strut and the plane goes down, but basically aerial hunting is an efficient way to reduce wolf populations. Skilled teams can kill a dozen or more wolves in a single day.

Aerial wolf hunting is still the centerpiece of Alaska's wolf-control work. Efforts to reduce wolves are undertaken primarily to increase game animals for hunters. As a 1979 Alaska Department of Fish and Game paper states: "The Department of Fish and Game acknowledges, as a basic proposition, that wolf-reduction programs which are intended to rehabilitate depressed ungulate populations are not needed to increase the population of either predator or prey species, but are for the sole purpose of providing more animals for human consumption."

When Alaska became a state, the game department took over management of the wolf from the federal government. In 1960 the department reduced wolf control in Game Management Unit 20, near Fairbanks, because the moose population there was large and relatively inaccessible to hunters. But over the next fifteen years, things changed. By the mid–1970s hunters were entering Game Management Unit 20A, a section of Unit 20, on motorized vehicles. Severe winters and the increased hunting—in 1973 hunters killed twice the average annual moose take of most of the previous decade—served to reduce the moose population from ten thousand in 1965 to twenty-nine hundred in 1974. The game department concluded, based on its own data, that wolves were a contributing factor in the decline and were keeping reduced moose numbers low.

The moose is the largest member of the deer family, but it looms even larger than life in the minds of many Alaskan hunters. The whole question of how wolf predation on moose affects hunting is fraught with heavy emotion, though at least one practical consideration underlies the question: For some hunters, moose meat is a staple of the diet, especially when beef prices soar as they did in the mid-1970s. Consequently, the hunting community was vocal in its support of wolf control. It also used its influence to ensure that the game department did not move to reduce the level of moose hunting. Instead, in February 1975, the department began the first state-managed aerial wolf hunt. However, Friends of the Earth, a conservation group headquartered in Washington, D.C., joined local conservationists in a court suit that stopped the hunt.

The following year the game department determined once again to hold aerial wolf hunts, this time in three game management units—20A, 13, and 5. In Unit 5, severe winters and possibly overhunting by humans had caused a decline in moose numbers. In Unit 13, the state wanted to experiment by wiping out all forty-five wolves in the 3,100-square-mile area to see how that would affect moose numbers. The task there was simplified because some of the wolves had been collared with small radios as part of a biological study. The radios transmitted signals that biologists could pick up on special receivers, allowing the wolves to be located at any time. All the aerial hunters had to do was tune in to the radio beeps, zero in on the packs, and start shooting. The goal in Unit 20A was to reduce wolves to the point that one wolf survived for every hundred moose.

Defenders of Wildlife, another Washington, D.C., conservation group, subsequently sued to stop the hunt. The hunt was delayed but eventually went ahead when Defenders lost in court. In Unit 13, all but two or three wolves were killed. In Unit 20A, 135 wolves were killed by state and private hunters, leaving perhaps 40 or 50. No wolves were killed in Unit 5 because a lack of funds stopped the hunt there. However, wolf control continued in Units 13 and 20A for the next two years. The wolf-to-moose ratio in Unit 20A eventually was brought to about one wolf for every forty moose. The game department wanted to bring the figure down to one wolf for every eighty moose, but even at the lower ratio the moose population started to increase. This fed the doubts of conservationists who believed the department's wolf-control program was poorly conceived and based on unsound data. In 1979, Defenders of Wildlife brought another unsuccessful suit to stop the hunt.

In 1986 the Alaska Department of Fish and Game was still attempting to initiate aerial-hunting programs that called for killing 80 to 90 percent of the wolves in some game management units. The department did this by tracking collared wolves with radio frequencies that were covered by a Federal Communications Commission permit. Defenders of Wildlife and the Animal Protection Institute in December 1984 succeeded in stopping this practice when they received a ruling from the Federal Communications Commission that use of the collars to hunt wolves was illegal. According to the commission, the permit authorized use of the radio-collar frequencies for scientific research only. However, the game department later circumvented this restriction by tracking the wolves on radio frequencies that were not covered by the original federal permit. Moreover, the Board of Game, a commission that sets game-department policy, has found another way to eradicate wolves. It has authorized a trapper education program in which trappers are given wolf-location data obtained from radio-collar studies and sold leghold traps and snares at wholesale prices. This trapper education program is even being conducted on an 18,000-square-mile area within Yukon Flats National Wildlife Refuge.

The killing of wolves from aircraft costs about $1,000 per wolf. In 1987 the Alaska game department gave up its hunt plans because it lacked the funds to carry them out. Nevertheless, the department still sees aerial wolf hunting as the best way to protect moose and caribou from wolf predation. This is not unreasonable, given the convincing scientific evidence which suggests that wolves can limit growth in prey populations already reduced by other factors, such as severe weather. But since wolves and their prey have coexisted for millennia, it is obvious that the prey species eventually would increase, if only because predation would decline as wolves themselves died from lack of prey. The issue in Alaska is whether the natural ebb and flow of wolves and their prey can be accommodated when priority is placed on the availability of moose and caribou for human sport. For those who would reduce wildlife to little more than a commodity for human entertainment, the answer is simple—control wolves. After all, control advocates say, the game department intends to wipe out only local wolf populations, a few hundred of the thousands in Alaska. This small loss will not hurt the total wolf population.

Those anxious about wolf control perhaps remember the words of Stanley P. Young, a wolf biologist and control proponent who wrote in 1946, "It is doubtful, however, whether the wolf will ever be completely extirpated from North America as it has been in most of Europe. It is felt that this interesting predator will long hold out in the wilder portions of Alaska, the Canadian Arctic, northern Mexico, certain areas of far-western Canada, and northern, western, and southern sections of the United States." Only twenty years after he penned those words, the wolf was in fact gone from the western and southern sections of the United States, nearly gone from the northern sections, and virtually extinct in northern Mexico. Those anxious about the future of the wolf see their options being irrevocably reduced in the United States to Minnesota and to Alaska.

On the other hand, the wolf is a resilient species. As long as a reservoir of wolves remains, the possibility exists that the animals can repopulate areas in which they have been wiped out. A good example is the Kenai Peninsula of south-central Alaska. Wolves were wiped out there in about 1915 by hunting, poisoning, and the burning of the woodlands during the gold-mining era.

The burning of the woods, however, was a boon for the moose, providing open areas with increased food plants. In the 1920s the Kenai moose reached high densities. To protect them, Franklin D. Roosevelt in 1941 signed an executive order creating Kenai National Moose Range. A fire in 1947 burned 590 square miles of forest, yielding more and better moose habitat. In the 1950s and early 1960s, moose populations peaked.

Some evidence suggests that at about this time wolves returned to the Kenai in scattered numbers—mostly lone individuals. Wolf hunting was subsequently closed in the area. Within fifteen years, the wolf population had increased to more than four hundred.

Unfortunately, at about that time the moose population crashed to about the pre-1947 level as a result of a series of severe winters and of damage to the habitat by an overabundance of moose. The giant deer, their populations unchecked, had eaten too much browse. Studies were undertaken to look at other factors in the moose decline. Researchers discovered that wolves in one Kenai study area were killing about 8 percent of the adult moose and 13 percent of the calves. Another study indicated that black bears were accounting for about 40 percent of the calves. In 1974, the Department of Fish and Game responded to this evidence by opening a sport-hunting season on wolves. It left black bears alone because they are popular game animals.

Although the U.S. Fish and Wildlife Service attempted to reduce the killing of wolves on the Kenai National Wildlife Refuge, in the late 1970s the state moved in the opposite direction. It increased the bag limit and the length of the season. Studies showed that the control effort was working: Radio-collared wolves lived an average of sixteen months after being collared, 88 percent of all wolf mortality recorded after six months of age was human related, and at least 42 percent of young adult wolves dispersing from their pack territories were killed. This last factor would reduce severely the ability of wolves to expand their range.

RETURN OF THE WOLVES

Experiences on the Kenai Peninsula show that if left alone wolves can repopulate good habitat from which they have been extirpated. With this in mind, some conservationists are urging the federal government to reintroduce wolves to suitable habitat in the lower forty-eight states. In one case, wolves have skipped the bureaucratic process and reintroduced themselves.

This happened back in 1982 in northwestern Montana, hard against the Canadian border, when a pair of Canadian wolves crossed into Glacier National Park. They subsequently bred and denned just north of the border. Two years later, biologists observed five to seven wolves as far as 25 miles south of the border. As the wolves matured, some moved into Glacier National Park and some into adjacent Flathead National Forest. Two wolf packs, including some twenty animals, now range back and forth across the U.S.–Canadian border.

The fate of these wolves is uncertain. Ranchers in the area, fearing loss of cattle, may shoot, shovel, and shut up—a colloquial way of saying they will kill wolves clandestinely. Human activity such as camping and hiking may frighten off wolves. And hunting in British Columbia may dim the chances of more coming into the United States. This last problem may be alleviated by the actions of the provincial government. In 1987 the province closed its wolf season after Montana wolves that had crossed the border were shot.

To help with the repopulation of wolves in Montana and Idaho, the U.S. Fish and Wildlife Service has appointed a special Northern Rocky Mountain Wolf Recovery Team. Recovery teams are panels of biologists, conservationists, and other interested parties, such as ranchers, who seek to develop plans for the restoration of endangered species. Such teams are supposed to be appointed for all species listed under the federal Endangered Species Act.

The wolf recovery team has laid out a plan that it hopes will be acceptable to human residents of the areas in which wolves are reestablishing themselves. The plan calls for three zones of wolf management. In the core of the wolf recovery area, wolves would be protected from human activities. The core would include such areas as Glacier National Park and perhaps parts of the national forests. Outside of the core zone would be an area in which wolf recovery is encouraged but in which problem animals—for example those that prey on livestock—can be killed. And in the third zone, made up primarily of agricultural and residential lands, wolves would be removed on sight. Wolf conservationists hope that the example of successful wolf management under Minnesota's similar plan will help garner local support for wolf recovery in Montana. If enough support can be found, and if wolves continue to filter quietly into Idaho and Montana, then perhaps the end of this century will find outdoor enthusiasts traveling to the northern Rockies to hear wolves howling there once again.

A more sensitive topic is the planned reintroduction of wolves to the Greater Yellowstone Ecosystem and parts of central Idaho. Both areas are too remote from the nearest inhabited wolf range for natural repopulation to occur, yet both have attributes that make them good candidates for reintroduction. Central Idaho has seven wilderness areas that would be suitable for wolves and that would keep them remote from livestock. The Greater Yellowstone Ecosystem includes Yellowstone National Park and adjacent wilderness areas and national forests. Wolves were once native to this area, but National Park Service control measures for the protection of moose and deer wiped them out in the 1940s. Between 1912 and 1926 alone, 130 adult wolves and 80 pups were killed there by the park service. Today, twenty-five thousand to thirty thousand elk as well as thousands of bison, moose, and mule deer—all natural wolf prey—inhabit the area. Elk populations, in fact, may even exceed the ability of the land to support them. Putting wolves back into the park might help reduce elk numbers and protect the park from the voracious herd.

Some ranchers and hunters in both areas oppose wolf reintroduction. In 1984 the head of the National Wool Growers Association, an organization that represents sheep ranchers, said that his organization opposed any plan to relocate wolves into the

Rocky Mountains. "This gives precedence to the predator over livestock interests," he said. "When the plan is released, we will take whatever action is necessary to stop it."

An opinion poll of attitudes about wolves suggests that an ingrained bias against predators exists at all potential reintroduction sites, principally among ranchers, farmers, and hunters. The Northern Rocky Mountain wolf recovery team has proposed two measures to accommodate this bias. First, any wolves released in the Rockies would be considered part of an experimental population, which in this case means that they can be trapped and killed if necessary. Second, a zoning management plan like the one for the Montana wolves would be initiated to delineate where the needs of livestock would take precedence over the needs of wolves.

Wolf biologists are convinced that wolves would be even less of a problem in the Rockies than they are in Minnesota, mainly because in Minnesota wolves and livestock are interspersed from at least May through October, while in the northern Rockies harsh winters would necessitate the removal of cattle from wolf range for all but three months of the year. Moreover, most grazing in the northern Rockies is along the periphery of the proposed wolf reintroduction area.

According to the wolf recovery plan for the northern Rockies, wolf restoration would be managed by the federal government until at least ten breeding pairs have been established for a minimum of three years in all three Rocky Mountain recovery areas—northern Montana and Idaho, the Greater Yellowstone Ecosystem, and central Idaho. At this point, wolf management would be turned over to the states, and the wolf would be removed from the endangered species list. Federal protection of the wolf would then cease unless the species again became endangered.

The National Park Service and the Forest Service are working with various wolf biologists and conservation groups, most notably the National Audubon Society, to determine when and how the northern Rockies reintroduction effort may take place. Few participants in the process are convinced that reintroduction will

108   ENDANGERED SPECIES

occur soon. A major stumbling block is the Montana Department of Fish, Wildlife, and Parks. The department not only has refused to join the federal government in a cooperative study of the wolves that are reoccupying the northern Rockies but also has opposed reintroduction plans. Instead, it wants wolf management placed in its own hands. Conservationists question the wisdom of acceding to this request. The state wildlife department has already said that it will not bankroll wolf recovery with funds from the sale of hunting licenses, the agency's primary source of funds. The money would instead have to come from the Montana legislature. But the legislature is dominated by anti-wolf elements. Besides, a legislature that has been unwilling to fund even education adequately is highly unlikely to fund wolf management. Congressional delegates from Montana and Wyoming have also opposed wolf reintroduction. Montana representative Ron Marlene has compared wolf reintroduction to releasing cockroaches in an attic.

Humankind has long set its mind to the destruction of the wolf. Knowing how thoroughly destructive humankind can be, it is impossible to say with conviction that the gray wolf will be with us always. But as long as people are willing to stand mutely in the darkness of a quiet summer night in patient anticipation of the wolf's wavering song, hope for the wolf survives.

CHAPTER 7

The
Grizzly
Bear

The grizzly rivals the polar bear as the most formidable predator on the entire continent of North America. Mature grizzlies can outweigh a baby grand piano. They are powerful enough to carry that bulk at a speed that can overtake a deer. And they have the aggressive personality of a heavyweight boxer. So threatening a beast presumably would loom large in the accounts of the first explorers to enter its domain, which was virtually the whole of North America west of the Mississippi, from Alaska to northern Mexico. Yet the records left by the first Europeans to penetrate the West do not speak loudly of the grizzly. They tend instead to whisper of it. The sixteenth- and seventeenth-century Spaniards who rambled in the Southwest and the Great Plains, presumably interested more in gold than in wildlife, noted only that they saw bears. The grizzly was not mentioned specifically until 1691, when Englishman Henry Kelsey saw the bear during his discovery of the Canadian prairies. In a journal entry dated August 20, he wrote of seeing "a great sort of Bear wch is Bigger than any white Bear & is Neither White nor Black But silver hair'd like our English Rabbit."

The few scant mentions of grizzlies left by the eighteenth century are vague. They describe a western bear light in color, massive in size, and vast in ferocity. This new bear was not lifted from obscurity until 1805, when the Lewis and Clark expedition first encountered the grizzly while boating through northeastern Montana. The explorers finally introduced it, via their journals, to the native inhabitants of civilization.

Here is what Lewis and Clark wanted their own and future generations to know about the grizzly:

Clark: May 5, 1805:

> . . . in the evening we saw a Brown or Grizley beare on a sand beech, I went with one man Geo Drewyer & Killed the bear, which was verry large and a turrible looking animal, which we found verry hard to kill we Shot ten Balls into him before we killed him, & 5 of those Balls through his lights This animal is the largest of the carnivorous kind I ever saw.

Lewis: May 5, 1805:

> it was a most tremendous looking anamal, and extreemly hard to kill notwithstanding we had five balls through his lungs and five others in various parts he swam more than half the distance across the river to a sandbar, & it was at least twenty minutes before he died . . . he measured 8. Feet 7½ Inches from the nose to the extremety of the hind feet, 5 F. 10½ Ins. arround the breast 1 F. 11. I. arround the middle of the arm, & 3.F. 11.I. arround the neck; his tallons which were five in number on each foot were 4⅜ Inches in length . . . this bear differs from the common black bear in several respects; it's tallons are much longer and more blont, it's tale shorter, it's hair which is of reddish or bey brown, is longer thicker and finer than that of the black bear; his liver lungs and heart are much larger even in proportion with his size; the heart particularly was as large as that of a large Ox. his maw was also ten times the size of black bear, and was filled with flesh and fish.

Lewis: May 6, 1805:

> saw a brown bear swim the river above us, he disappeared before we can get in reach of him; I find that the curiosity of our party is pretty well satisfyed with rispect to this animal, the formidable appearance of the male bear killed on the 5th added

to the difficulty with which they die when even shot through the vital parts, has staggered the resolution [of] several of them, others however seem keen for action with the bear.

Lewis: May 11, 1805:

The wonderful power of life which these animals possess renders them dreadful; their very track in the mud or sand, which we have sometimes found 11 inches long and 7¼ wide, exclusive of the talons, is alarming; and we would rather encounter two Indians than meet a single brown bear.

The expedition continued to encounter and shoot more grizzlies, which they also called brown and white bears. They also encountered Native Americans, some of whom made it clear to Lewis and Clark that the white, reddish, and dark brown bears with the long, white claws were all of one type, and distinct from the smaller, shorter-clawed black bears. Also, the Natives said, the grizzlies usually do not climb trees, while the black bears do. Hides and bones sent East by the expedition revealed this distinction to the scientific world, giving the grizzly its first recognition as a species.

The repeating rifle, the railroad, and the cattleman began pushing the grizzly off the plains in the last half of the nineteenth century, but still the animals were plentiful in many areas. In 1877 one visitor to Montana's Big Horn Mountains saw nine grizzlies in one day. Hunters in 1883 spotted eleven of them outside Spokane, Washington, during a single weekend. In 1885 two men traveling the foothills southwest of Bakersfield, California, counted fourteen grizzlies in one spot. During a trip to Idaho's Bitterroot Mountains in September 1902, naturalist Ernest Thompson Seton saw a half-dozen grizzlies in a single month. But by then the grizzly was on the wane. Only twenty years later, mammalogist C. Hart Merriam would report that fewer than a thousand remained in the lower forty-eight states. The grizzly, like the wolf, had fallen to the inroads of settlement. In 1926 Seton predicted that within five years the grizzly would survive only in Yellowstone National Park.

By the late Twenties the grizzly had lost ground throughout its range. It vanished from Texas in 1890. In 1917, J. Stokely Ligon, who masterminded the extermination of the wolf in the Southwest, reported that only forty-eight grizzlies remained in New Mexico. The following year he announced that he was making plans for an extensive poisoning program. By the late 1920s Ligon's efforts had paid off, and he was becoming concerned that all grizzlies in the Southwest would soon be killed off. He was also becoming skeptical about the ranchers' approach to bear control. He came to believe that certain individual bears learned to kill livestock and that these animals should be removed. But ranchers, he asserted, used the killing of a cow by a bear to justify the slaughter of all grizzlies. Ligon also concluded that in many cases grizzlies turned to livestock killing because the ranchers had put too many livestock on the range. The cattle badly overgrazed the land, wiping out the vegetable matter that made up a large part of the grizzly's diet. Hungry bears were forced to feed on livestock.

In 1927, the New Mexico legislature declared all bears protected game animals but allowed ranchers to obtain permits to kill bears that preyed on livestock. Although the federal control program abated, ranchers persisted in killing grizzlies into the 1930s, using poisons and hiring hunters and hounds to track down and kill individual bears. By the mid–1930s the last grizzlies in Utah, New Mexico, and Arizona had been slain.

California's last grizzlies—including a golden-haired coastal breed that sometimes fed on the beached carcasses of whales and sea lions—disappeared in 1922. By the 1950s only a handful of grizzlies remained in the lower forty-eight states, primarily on protected areas such as national parks. A few hung on in northern Mexico until the U.S. Fish and Wildlife Service supplied Mexican ranchers with Compound 1080. Used indiscriminately against predators, the poison probably helped wipe out the last of Mexico's grizzlies in the early 1960s. David E. Brown, a wildlife biologist who chronicled the decline of the grizzly in his book *The Grizzly of the Southwest*, searched for the species in its Mexican range in 1982 but found none. He was told no grizzlies had been observed there in twenty years.

The extirpation of U.S. grizzlies south of Yellowstone National Park seemed a *fait accompli* as early as 1954, when a grizzly in Saguache County, Colorado, was killed. This was the first grizzly killed in the state since September 1951, when a U.S. Fish and Wildlife Service trapper killed a two-year-old boar west of Creede. It was the last grizzly killed for twenty-five years. Long before that time had passed, biologists had presumed that the grizzly was gone from Colorado. Then, on September 23, 1979, a 375-pound female bear wandered through the San Juan National Forest in southwestern Colorado and came to the head of the Navajo River northwest of Blue Lake. That evening, Mike Niederee, an archery hunter from Kansas who was looking for elk, came to the same spot. His guide was Ed Wiseman of Crestone, Colorado. The story Wiseman and Niederee told about what happened that day has raised some eyebrows.

According to the story, Niederee was about 10 or 20 yards from the bear when he spotted it. Apparently it was sleeping but soon became aware of Niederee and ran off. It had gone several hundred yards when it ran into Wiseman and attacked him, biting his right shoulder and legs. When Wiseman's attempts to play dead failed to stave off the bear's attack, he stabbed it in the throat and neck with a hunting arrow. The bear retreated a short distance and died.

Niederee had not seen the attack, but he did hear Wiseman's cries and came to his aid. Wiseman was badly wounded, and Niederee went for help. A helicopter picked up Wiseman the next morning. He spent several weeks recuperating.

Three days after the attack, the bear's hide and skull were flown out by helicopter. When state biologists got a look at the remains they realized that the bear was a grizzly, the first identified in the state since 1954. They retrieved the rest of the body. Evidence showed that the bear had died of a wound that penetrated its body through the rib cage and severed the aortic arch, a large vessel of the heart. This evidence cast some doubt on the story the men told, since it was unlikely that the wound could have been inflicted with a hand-held arrow. Officials accepted the story, however, and no charges were brought against Wiseman or Niederee.

The body of the bear was too poorly preserved to show whether the sow had ever bred, which would have been the surest evidence to show whether she was a lone bear, the last of her kind in the state. Nevertheless, the bear's presence stimulated the state wildlife agency to undertake a two-year search for signs of grizzlies in the San Juan Mountains. Nothing ever turned up, and the study ended. Thus was the record closed on what was quite likely the last grizzly in Colorado, if not the last south of Yellowstone National Park.

Seton's prediction that the grizzly would survive in the lower forty-eight states only in Yellowstone has very nearly come true in this generation. About two hundred remain in the park and adjacent areas. Another five hundred to six hundred live in and around Glacier National Park and national forest wilderness areas in northwest Montana. These animals are part of a population whose range extends into Canada. A handful is thought to survive in the North Cascade Mountains of Washington State, but the last verified Cascades grizzly was killed by a hunter in 1967. A small population might survive in the Selway–Bitterroot Mountains on the eastern border of Idaho and Montana. Reports of grizzlies come in from this area every year, but no sightings have been confirmed since 1946. A small number live in the Selkirk Mountains of Idaho and British Columbia. Another dozen or so live a bit west of Glacier in the Cabinet Mountains–Yaak River drainage area. Biologists believe the Cabinet population will not survive unless bears from stable populations are captured and released in the Cabinet portion of the area. The tenuous status of these bears—caused by loss of habitat and uncontrolled killing by people—puts them more in the category of endangered than of threatened animals, though this has never been officially recognized. However, the U.S. Fish and Wildlife Service has developed a plan to release grizzlies into the area. The plan originally called for releasing four to eight bears as a test of the technique, but local opposition caused the service to reduce the figure to only four female grizzlies by 1990.

In all, the grizzly of the lower forty-eight states survives on about 1 percent of its former range, at about 1 percent of its pristine numbers. In 1975 grizzlies in the lower forty-eight states were listed under the Endangered Species Act as threatened. Seven years later the U.S. Fish and Wildlife Service completed a grizzly recovery plan. The plan's stated goal is to establish viable populations wherever the grizzly occurred in 1975. A special committee formed in 1983 oversees grizzly management. Called the Interagency Grizzly Bear Committee, it is made up of representatives of the National Park Service, Forest Service, Fish and Wildlife Service, Bureau of Land Management, Bureau of Indian Affairs, and the Idaho, Washington, Montana, and Wyoming wildlife agencies as well as Native American tribal representatives and officials from Alberta and British Columbia. It attempts to coordinate and implement the management and research activities of all involved agencies, giving it a crucial role.

THE GRIZZLY TODAY: A LAST STAND?

Most of the 20,000 square miles on which grizzlies still wander in the lower forty-eight states are managed by the U.S. Forest Service as national forests. These lands are among the grizzly's last havens against the onslaught of civilization. But the national forests are not entirely safe. The major problem is that the forests are multiple-use lands. Consequently, administrators are required by law to manage the forests for a number of sometimes conflicting uses, such as mining, wildlife conservation, sport hunting, camping, livestock grazing, and oil and gas development. Federally designated wilderness areas offer somewhat better protection. No permanent developments are supposed to be built on wilderness areas, and only footpaths are allowed. But even in the wilderness areas some commercial activities, such as oil and gas development, can take place if, for example, an oil company had a permit to explore and drill for oil on the land before it was designated as a wilderness. The development, in legislative parlance, is grandfathered into the congressional bill that makes the area a designated wilderness. In most cases, developers are given a deadline for development. Once the deadline passes, their grandfather clause is finished.

Primary among the many uses of the national forests is timber production. The Forest Service offers tracts of land for logging, and timber companies bid for the right to cut them. To make the cutting easier, roads are laid in the forests. Usually built at Forest Service expense, roads cost so much that more than a third of federal timber sales lose money.

The roads have a negative effect on grizzlies, too. Research shows that some grizzlies do not adapt well to the influx of roads and human activity. They move out of their chosen habitat, forced to live on smaller and smaller patches of land. Bears that do adapt face a different set of problems. The human activities that roads bring increase the chance that grizzlies will have fatal confrontations, since people are the cause of the vast majority of grizzly deaths in the lower forty-eight states.

Logging can benefit grizzlies by creating openings in which prime bear foods grow. The Forest Service has launched studies of grizzly habitat use to determine the type of cuts that will be most beneficial. However, the service persists in scheduling massive cuts that jeopardize grizzlies by devastating large tracts of habitat. Conservationists thus keep a wary eye on the Forest Service and its logging plans.

Another land use that has conflicted with grizzly conservation is livestock grazing. Something in excess of 100 million acres, or slightly more than half the National Forest System, are covered under grazing permits. This means that grizzlies are forced to share parts of their last stronghold with cattle, sheep, and horses. Until 1975 this was a particularly bad situation because ranchers were permitted to shoot bears that threatened livestock.

Because the grizzly could not long survive if ranchers were given a free hand with their rifles, conservationists fought hard against the Forest Service's grazing practices. They argued that grizzlies in the national forests should be given priority over cattle and sheep, that livestock should be removed from grizzly range. Ranchers countered that grazing these areas was an old tradition. In some cases, the ranchers pointed out, their permits covered national forest lands that had belonged to their direct ancestors years before.

Eventually, the conservationists scored a victory. Forest Service policy now requires that ranchers remove their animals from any pasture in which grizzlies and livestock conflict. Private conservation groups have offered to compensate ranchers for lost livestock, but most ranchers have turned them down. Apparently, they are offended by what they see as a handout and prefer to deal with the bears themselves. Nevertheless, tighter controls have helped grazing continue without diminishing the grizzly's chances of survival.

National forests are open to oil and gas development and mining as well as logging and grazing. The explosions used in seismographic oil exploration may cause some bears to abandon their range. Oil, gas, and mineral exploitation also bring with them the same problems that accompany logging—roads and increased numbers of people. Grizzly biologists and other conservationists monitor oil and mining activities closely and often with alarm. For example, in early 1983 ASARCO and U.S. Borax were planning exploration for minerals in the Cabinet Mountains Wilderness Area. This wilderness contained a small, vulnerable grizzly population. Plans also were being slated at the same time to build a ski resort two miles east of the wilderness and only five miles from the site of the mineral exploration. Consequently, bears were beset with the cumulative effects of several developmental activities. The U.S. Fish and Wildlife Service successfully demanded that the potential effects of the activities be studied before any actions were taken.

In some cases, the careful attention conservationists give the grizzly has hardened local opposition to bear protection. Developers who fear their plans will be stymied in favor of the bear use political influence to cripple conservation measures. For example, timber companies have told employees that grizzly protection will take away their jobs by placing limits on logging. The companies have urged employees to write letters to their congressmen opposing grizzly conservation. Such letters find willing readers. In 1984 and 1985, Idaho senator James McClure amended federal appropriations bills to slow plans to augment the Cabinet–Yaak bear population with grizzlies relocated from other areas.

National parks, unlike national forests, are supposed to be kept in nearly pristine condition. Most are closed to hunting, and all are closed to oil and mineral development. Despite the seemingly foolproof protection that parks appear to give to wildlife, some of the most explosive battles over the grizzly have been fought in the parks. Foremost among these battlesites is the one called Yellowstone.

Yellowstone National Park was created in 1872. It was the world's first national park and was to be preserved as a recreational area for the people. It was not then the inviolate wildlife refuge that it is today. The U.S. cavalry had to be called in to keep hunters from wiping out Yellowstone's bison, virtually the last wild bison in the lower forty-eight states. For a while the military was even put in charge of the park. Such vagaries of park management ended in 1916, when the National Park Service was created to administer the nation's parks.

Yellowstone National Park lies at the center of the Greater Yellowstone Ecosystem, the southernmost area known to be occupied by grizzlies today. The ecosystem includes Yellowstone and Grand Teton national parks, seven national forests, and three national wildlife refuges. At 14 million acres it is the largest primarily pristine area in the lower forty-eight states. Administration of the area, however, has at times made it less than ideal for grizzlies. The most infamous case involved the closing of park garbage dumps, where many bears fed.

The dump closures were rooted in the early 1960s, when a special advisory committee was appointed to study the management of the national parks. In 1963 the committee produced what became known as the Leopold report after committee chairman A. Starker Leopold, Aldo Leopold's son. In it the committee urged that the parks be managed with as little human interference as possible in an effort to make them as pristine as they were the day Columbus arrived on New World shores. The National Park Service concluded that for the grizzlies, pristine meant no garbage dumps. In 1969, the service began closing the dumps. This evoked an immediate response from two grizzly biologists, Frank and John Craighead.

The Craighead brothers had begun research on park grizzlies in 1959. The first major study of a grizzly population, their work was considerably enhanced by the use of radio collars, a new technology. These collars, put round the bears' necks, carried small radio transmitters that broadcast signals that could be picked up on special receivers. Each collar broadcast a different frequency, permitting the Craigheads to locate and track individual bears. They discovered that the bears depended heavily on garbage dumps as a food source. The brothers even suspected that many Yellowstone grizzlies were completely dependent on the dumps.

The Craigheads recommended that the park close the dumps gradually, weaning the bears from that source of food. The park service instead closed them abruptly. Forced to look for new food sources, grizzlies switched to eating in the dumps of nearby towns and rummaging through small towns' dumpsters. They also continued to invade campsites. Some motel owners compounded the problem by putting food out to attract bears for their customers. The park service responded by trapping the bears and shipping them 50 or 60 miles into the mountains, the backcountry. They could not be moved farther or they would have been taken out of the ecosystem. The transported grizzlies, however, soon returned to their home ranges. The service then started killing grizzlies on the presumption that garbage-addicted animals would be exterminated, leaving the park to grizzlies that lived full-time in the backcountry on natural foods. The Craigheads protested. They argued that *most* park bears were garbage bears and that killing them off would be tantamount to wiping out the bulk of Yellowstone's grizzlies. The service ignored their advice and in five years killed roughly half the park's population, according to Craighead estimates. In 1971 the Craigheads and the park service conflicted over research goals, and in 1971 the service refused to renew the Craigheads' contract.

For more than a decade after the Craigheads' era, Yellowstone grizzly numbers remained low. Grizzlies were thought to be about 30 percent below the minimum number that the recovery plan declared necessary for a viable population. The problem was compounded because only about a quarter of the animals were females. For a species that breeds slowly—sows usually have only two cubs every two or three years—an increase to the three hundred called for in the recovery plan seemed unlikely. In 1982 Roland Wauer, chairman of the Interagency Grizzly Bear Steering Committee, forerunner of the Interagency Grizzly Bear Committee, sent a memo to committee members in which he wrote, "Evidence persists that the population of grizzly bears within the greater Yellowstone ecosystem has seriously declined in recent years. *Unless some change occurs to reduce the grizzly's mortality rate soon, the probability of retaining this wildland species in Yellowstone National Park is minimal.*" This warning alarmed the conservation community and helped force a response from the bureaucrats who ultimately direct the course of grizzly management.

Why the Yellowstone grizzlies tottered so precipitously on the brink was a major question. Studies indicated that—dumps or no dumps—food was not the problem. But an obvious clue to the bear's trouble was this: Most grizzly deaths were caused by humans. Of thirty-four grizzlies under biological study that died from 1975 to 1983, three died from natural causes and one from unknown causes. The other thirty deaths were caused by humans. Some bears were killed by poachers, others by the park service itself because the animals were problem bears.

Problem bears are those that get in the way of humans by wandering into towns, invading dumps, raiding campgrounds, or killing livestock. The federal agencies involved in bear management have tried to deal tolerantly with these animals. In recent years some bears, particularly sows, were trapped and relocated five, six, seven times. At some point, however, the agencies would give up on individual bears and either kill them or turn them over to a research facility or zoo. Either way the bear was dead as far as the Yellowstone Ecosystem was concerned, since it would never reproduce there.

Ironically, some actions taken by the park service have helped to create problem bears. One example in recent years was the approval given by the park service for construction in Yellowstone of Grant Village, a mammoth complex that includes lodging units for nearly three thousand people, along with parking lots, recreational facilities, restaurants, a sewage treatment plant, a post office, and several other facilities. It was built on the confluence of several trout streams, a place that until the construction of Grant Village was a primary grizzly feeding site. Doubtless bears returning to the area in search of nourishment will come too close to people and will be dispatched as problem animals. Grant Village was completed in 1983 with the promise that when it opened, the park service would close Fishing Bridge, another recreational facility built at a grizzly fishing site. This was considered an important step, since nearly half the problem grizzlies killed or removed from the Yellowstone area first became problem bears because of habits picked up at Fishing Bridge. In a 1984 report, the park service itself called Fishing Bridge "an ecological mistake with serious consequences." The service since then has sought to close Fishing Bridge, but its efforts have been blocked by the Wyoming congressional delegation. The delegation wants the recreation area left open because it generates county tax revenues. So far, the park service has succeeded in closing a Fishing Bridge campground, but a recreational-vehicle park has been kept open.

Despite such problems as these, human-caused bear mortality has been reduced and grizzly protection has been improved greatly since Wauer made his dire prediction. The agencies involved in grizzly management have undertaken educational programs that show Yellowstone-area residents and visitors how to bear-proof campsites, garbage dumps, and hunting camps, thus reducing the number of animals that become problems. Since 1982, bear removals by federal agencies have declined. From 1982 through 1987, official removals accounted for an average of only about 1.3 bears yearly. Revisions being made to the grizzly recovery plan will limit all human-caused bear deaths to an average of no more than two females per year over any six-year period. This means that no problem bears will be removed if the limit is exceeded by deaths caused, for example, by automobiles or poaching. The new policy also will help ensure that the old days of excessive problem-bear removals will not return.

Yellowstone Ecosystem bears seem to be responding well to the improved management. In 1988, the lowest estimate for the population rose from 140 to nearly 170 bears. The head of the federal recovery effort, Chris Servheen, says that more females were observed with cubs in 1988 than in any previous year. Moreover, the fire that burned over large portions of Yellowstone National Park in summer 1988 may prove a boon for grizzlies. Some three hundred elk were killed, providing grizzlies with bountiful food. In 1989, vegetation growing on the burned sites should give bears another plentiful food supply. Elk, too, will benefit from the new growth, perhaps adding more elk meat to the bear's cupboard. The increased food supply should give a boost to grizzly breeding.

Servheen believes that the bear population will soon meet the recovery plan goal, after which it can be removed from the endangered species list. This is the surest measure of a recovery program's success. However, Servheen expects conservationists to oppose delisting. He predicts that many conservationists, recalling the Yellowstone grizzly's abysmal status only a few years ago, will fear that delisting will lead to management neglect and another decline. Servheen says that a carefully designed conservation plan for continued attentive management will be needed both to allay conservationist fears and to ensure sound grizzly protection.

Controversy over the delisting of a grizzly population is already in the air in a portion of bear habitat that wildlife managers call the Northern Continental Divide Ecosystem. This area in northwest Montana includes 1,600-square-mile Glacier National Park. In addition, the ecosystem includes 1,800 square miles of designated wilderness managed by the Forest Service plus some 4,000 square miles of national forest lands. The area is fraught with the usual problems. Located on the eastern edge of the Rockies, it faces increased development. The lower areas are rapidly becoming prime real estate. Logging and oil and gas development are under way. In addition, the divide is the only area in the lower forty-eight states in which grizzlies are hunted for sport.

The hunt is managed by the Montana Department of Fish, Wildlife, and Parks. Every year the department sells about six hundred special permits entitling the holder to shoot one grizzly bear during a fall hunting season. The state allows no more than fourteen bears, including no more than six females, whichever comes first, to be killed yearly from all causes. This means that if during the spring and summer six bears are killed to control livestock damage, then hunters may kill no more than eight others. The hunt can be stopped on forty-eight hours' notice when either the overall quota or the quota for females is reached. In recent years hunts have been canceled under these terms.

An effort to stop the hunt was launched in 1984 by Defenders of Wildlife. Their impetus came from winning a 1983 lawsuit over wolf protection in Minnesota. In that suit, Defenders had sought to block a plan by the federal Department of the Interior to turn over wolf management to the Minnesota Department of Natural Resources. The state, Defenders had claimed, planned to open a sport-hunting season on the wolves, which are listed as threatened. Defenders had argued successfully that the federal government could not simply turn over to a state its responsibilities under the Endangered Species Act and that it was illegal to hunt a threatened species unless a local population had grown excessive and would benefit from the hunt.

After winning the case, Defenders turned its sights on the Montana grizzly hunt. The grizzly is a threatened species, and the state had no scientifically sound data on the hunted population proving that the bears would benefit from hunting. Defenders threatened to sue the state and the U.S. Fish and Wildlife Service for permitting the hunt. However, Defenders main goal was not to stop the hunt, but to ensure that the state was monitoring the bears carefully, basing the hunt on sound data, and capable of supporting the hunt with biological facts. Consequently, Defenders accepted a promise from the state wildlife department that the agency would undertake a study of the grizzly population to be sure the bears' numbers would not be harmed by hunting. After the state completed its study, it reduced the overall quota on bears from twenty-five to the current fourteen and initiated the subquota on females.

One outcome of the Defenders' action may be a petition from the state of Montana asking the Fish and Wildlife Service to delist the grizzly. After conducting its research, the state concluded that the divide ecosystem has 549 to 813 grizzlies. The recovery plan goal for the ecosystem is 560. The state argues that recovery has been accomplished, so delisting should occur.

Some grizzly biologists disagree. They say that the state is using imprecise figures that originated in research that was not meant to determine population numbers. They fear that the state wants the bears of the divide ecosystem delisted so that it can take control of the species' management. This concerns conservationists because the state legislature is dominated by livestock interests. State management of the divide-ecosystem bears may put the animals in the hands of people who would prefer to shoot them wherever livestock is threatened, even in national forests. Further, it may put the purse strings of grizzly management in the control of people opposed to bear conservation.

Chris Servheen says that in any event, data must be collected for at least two more years before a decision about delisting the Northern Continental Divide grizzly can be made. Data collected so far suggest that delisting may be appropriate, but, Servheen says, other factors also must be considered. As in the Yellowstone situation, delisting will have to await a sound conservation plan that describes clearly in black and white the constraints necessary under state management for continued protection of the grizzly and its habitat.

TOMORROW

The grizzly once roamed this continent with a rogue's prerogatives. Nothing could stand in its path. Today its numbers are strong only in Alaska and in remote parts of Canada. Even there the grizzly's survival seems more assured than it actually is. About 20 percent of Canada's grizzlies are in British Columbia, and about half those bears live in a mid-coastal region that is being heavily logged. The Yukon holds about 60 percent of Canada's grizzlies, and in the Yukon bears are blamed for killing the game animals that hunters want. Provincial officials have initiated aerial hunts, authorizing government hunters to shoot the bears from aircraft in an effort to reduce their numbers. In Alaska, the grizzly is itself a valued game animal, so it is not killed by airborne gunmen. But if the hunting community should ever conclude that grizzlies feed too greatly on moose or caribou, the bear, like the wolf, could become subject to control campaigns.

William T. Hornaday, a leading conservationist at the turn of the century, wrote in 1913, "A Rocky Mountain without a grizzly upon it, or at least a bear of some kind, is only half a mountain,—commonplace and tame." Survival for the grizzly bear depends on those who do not want an entire continent made commonplace and tame. Only if untroubled wilderness is eternally preserved will future generations see such living monuments of a pristine North America as the grizzly bear.

CHAPTER 8

The
California
Condor

The California condor has become a cause célèbre. It has assumed almost mythological proportions in the minds of many, even polarizing the more scientifically inclined. During the past two decades debate about condor protection has triggered some of the most violent disputes ever to shake the conservation community. The reverberations rattle windows even today. In the world of condor management every participant—whether scientist, amateur naturalist, rancher, forester, journalist, professional conservationist—shares one trait: All are ardent, even vehement, in their convictions about what should or should not be done to save the condor.

And what is this creature that has generated such profound emotions? It is a vulture, a *giant* vulture. It is mostly black, except for triangles of white beneath each wing. Its red-orange head is naked, as the heads of vultures always are. Its eyes are a glittering ruby red. It feeds on dead things, finding food at least in part by watching eagles and following them when they land to feed. It is a master at the craft of soaring. It can ride the wind for miles and hours with barely a flap of its wings.

Which among these traits has attracted the condor's following? Is it the spectre of a large black shadow aloft in the sky? Is it the sound of the wind whistling audibly through the feathers at the tips of its wings? Is it its beauty, for those gifted enough to see it? It may be all of these. But probably the condor's strongest attributes are its rarity and its size. For nearly a century the bird has hovered near extinction. Probably not a hundred condors have survived at any one time in this century, and it has seemed for decades that it would vanish altogether at any moment. And this would be a singularly terrible loss, since the condor is the largest bird in North America and one of the largest flying birds in the world. It would be a major tragedy for those who fear losing superlative creatures: the fastest, the heaviest, the tallest. Several years ago a scientist cut down what was thought to be the oldest tree in North America just so he could count its growth rings. Such ignorant behavior on the part of a presumably educated man disturbs a lot of people. They are similarly disturbed by the thought of losing the continent's largest bird.

Amazingly enough, for all the interest the condor has generated, very little is known about it. If all the definitive texts covering everything known about the bird were stacked atop one another, the pile would tower all of three inches high. More is known about distant stars and galaxies than about this massive bird that soared the skies for millennia and ceased to exist in the wild at 10:15 A.M. on Easter Sunday 1987.

A LONG AND LINGERING DEATH

Eleven thousand years ago the bird that came to be called the California condor ranged widely over North America. The bones most commonly found in the rich fossil beds of southern California's Le Brea tar pits are those of the condor. Its remains have been discovered in Nevada, New Mexico, Texas, and Florida, and bones found in 1987 in New York State confirmed that it ranged clear into the pine woods of the Northeast. It fed, biologists think, on the remains of massive Ice Age mammals—woolly rhinos and mammoths, giant sloths and giant beavers. When the Ice Age and its glaciers faded into the north, these creatures vanished and the condor's range began to ebb. Apparently, the extinction of the large mammals critically reduced the condor's food supply. This much is speculation. What is known for sure is that the condor was restricted almost entirely to the West Coast by the time the first Europeans arrived in the New World. Why it should survive there when it had disappeared from the rest of its range is uncertain, though it may have had something to do with the availability of large food sources on the West Coast. In spring the rivers of the Northwest Pacific were clogged with dying salmon. The beached carcasses of whales and sea lions may have provided additional sources. But why did the condor fail in the East when it had wood bison to eat? And weren't whales as likely to wash up on the Atlantic Coast as on the Pacific?

The first Europeans to see the California condor sailed aboard three small ships into Monterey Bay in December 1602. They had left Acapulco seven months before to explore the Pacific coast. From the deck of one ship a Carmelite friar gazed upon the carcass of a whale at the edge of the shore. At night, grizzly bears came to feed upon it. By day huge black vultures squabbled over it. In his journal the friar wrote that the birds resembled turkeys. He declared that they were the largest birds he had seen on the voyage and guessed their wingspans at more than 11 feet. Though he overestimated the wingspan of even the largest condor by at least a foot, the rest of his description fit.

After that early expedition the coast of California remained virtually unexplored for over 150 years. Then, in 1769, the condor reappeared in the annals of European history, encountered frequently by a Spanish expedition traveling over land in search of Monterey Bay. Near today's Watsonville, California, Native Americans showed the explorers the carcass of a condor they had used in a ceremony. Despite this contact, the first specimens did not reach Europe until 1792. These were two stuffed birds. One apparently was lost soon upon its arrival in the Old World. The other, kept in England, was damaged, making it a poor example. But since it was the only North American condor known to science until late in the 1820s, it was highly valued. Naturalists traveled from the continent just to look at it.

The members of the Lewis and Clark expedition were the first Americans to become familiar with the condor. They shot several in the vicinity of the Columbia River, in Washington State, but all they are known to have brought back to civilization was a skull and a wing feather. Scientists did not have enough material to classify the bird successfully until the early 1830s, and that feat was accomplished not in America but in Europe.

Ironically, the condor no sooner was added to the world's inventory of species than its numbers began to drop. In an 1839 edition of *Ornithology in The United States*, naturalist J. K. Townsend followed up a trip to the Columbia River by writing, "The California Vulture cannot, however, be called a plentiful species, as even in the situations mentioned it is rare to see more than two or three at a time. . . ." He was perhaps more correct in his observation than he realized. The last known condor of the Columbia River region, where Lewis and Clark had become familiar with the birds, was killed in 1834—by J. K. Townsend.

By the 1840s condor range was primarily, if not entirely, confined to California. Apparently, the bird was fairly common there. California hunters complained that condors pilfered any game left unguarded. A doctor who traveled up the Sacramento Valley in 1855 as part of a railroad survey crew recorded seeing condors soaring overhead every day. But the bird was clearly on the wane. By the 1870s it was rare around Monterey. By 1890 it had vanished from northern California. That year James G. Cooper, a noted ornithologist in his day, sounded the first alarm in behalf of the condor, writing that the bird was "in the process of extinction."

As the condor became increasingly rare, it also became increasingly sought by amateur collectors and museums. Taking eggs and birds for specimens was a major factor in the condor decline. Sport hunters also took a heavy toll. From 1900 to 1920, its range in the southern part of the state receded. A population found only in a single mountain range in Baja California vanished sometime around 1930. After that, the condor survived only in a horseshoe-shaped band that traced the southern end of the San Joaquin Valley. There, among dry, rugged mountains and tangled woodlands, the bird made its last stand.

THE CALIFORNIA CONDOR MEETS MODERN SCIENCE

The first serious study of California condors, begun at a nest near Pasadena in the San Gabriel Mountains, was primarily an effort to photograph the birds. Given that the study was undertaken in 1906, the dedication and tenacity of the two men involved can scarcely be imagined. Hiking through grim, rocky terrain with cumbersome cameras that burned images on heavy plates of glass, William Finley and Herman Bohlman, of Portland, Oregon, collected the first visual record of wild condors. Their work stood the test of time, remaining for many years the best material on the birds.

After Finley and Bohlman, little attention was paid the condor until the 1930s. But the Thirties proved a crucial decade. First, in 1933 a forester at Los Padres National Forest, a focal point for condor activity, started a five-year study of the birds. Though he had to base his work on such shaky data as random observations of condors from fire towers, the forester was able to conclude with some reliability that no more than seventy of the birds survived. He also discovered that condors concentrated much of their activity in the area surrounding Sisquoc Falls in Santa Barbara County. This led the U.S. Forest Service to establish a 1,200-acre condor sanctuary around the falls in 1939. Ironically, the condors moved out shortly afterward.

The Los Padres study helped kick off the next major development in condor research. It began as a proposal by John Pemberton, a Los Angeles geologist, to collect motion pictures of condors. He took his idea to Joseph Grinnell, director of the University of California's Museum of Vertebrate Zoology in Berkeley. He also suggested that Grinnell assign a graduate student to undertake research on condor behavior. Grinnell chose a young man whose name eventually would become synonymous with condor studies: Carl Koford.

Koford, a zoologist who formerly had studied small mammals, entered condor range in the spring of 1939. In 1941 his work was interrupted by World War II and a stint in the navy. He returned to condor research at war's end, determined to come up with a good estimate of the condor population and its health. The National Audubon Society, which had funded Koford's research, published the results of his work in 1953. The 154-page document represented virtually everything known about the condor up to that time. But even before the report was finished, Koford's data had alerted various agencies and organizations to the condor's plight and had helped produce results. For example, in light of Koford's work, the Forest Service in 1947 established the 35,000-acre Sespe Wildlife Area, renamed the Sespe Condor Sanctuary in 1966. A year later, a technical group, the California Condor Advisory Committee, was established to advise personnel at Los Padres National Forest about condor management. In 1949 the International Technical Conference on the Protection of Nature met in Lake Success, New York, and numbered the California condor among thirteen birds throughout the world in need of emergency help. And in 1951 a part-time warden was assigned to the Sespe sanctuary, expanded that year to 53,000 acres. The warden's salary was covered jointly by the Forest Service and the National Audubon Society.

Publication of Koford's work was a major landmark in condor history. After years of field work, Koford was able to provide the first reliable accounts of condor behavior. However, because condors live in rugged country, range as much as a hundred miles in a single day, and usually nest in caves and on ledges on the sides of sheer, rocky cliffs, even Koford's accounts were sketchy. Perhaps Koford's most important accomplishment was a reasonably scientific estimate of condor numbers. He concluded that only about sixty survived and that there was nothing to indicate that the population was growing. Convinced that the bird could not survive much longer without human help, Koford made a series of recommendations for condor management. He called for better legal protection, including development of federal protection; a ban on public entry to federal lands used by condors; an educational program that would give people a better understanding of condor needs; and an effort to persuade ranchers to leave livestock carcasses on the range for condor food.

Much of what Koford sought was simply an enlargement of measures already being taken. For example, although no federal laws protected condors in 1953, some state laws did. Also, the Sisquoc River area had been closed to deer hunters since 1941, and the Department of the Interior had banned mining and mineral leases on 55 square miles of the Los Padres National Forest. Moreover, in response to Koford's study, the Forest Service in 1947 had prohibited public use of a large area along the lower Sespe River. Koford thus did not seem intent upon blazing new ground for condor protection. He even argued against three potentially important proposals. One was feeding the birds artificially, which he said was impractical. The second was transplanting the birds to areas they had formerly inhabited, which Koford considered impossible because the birds range so widely. The third was captive breeding, which he felt would reduce the condor to something uninspiring, pitiful, and ugly. Oddly enough, despite Koford's arguments, every one of these measures would be adopted in the not-too-distant future. But before they were, stormy battles would be waged over them within the environmental community.

THE CALIFORNIA CONDOR WARS

Concern for the condor began to mount preciptitously in December 1964, when a National Audubon Society study determined that the condor population was down to about forty birds. This was particularly alarming because Koford had shown conclusively only ten years before that the population had been stable at sixty birds for at least thirty years. The society responded promptly to this evidence of a sudden decline by appointing a full-time staff member as condor naturalist. His job was to teach people about condors, particularly how to identify them so that they would not be shot. He also was assigned to compile a list of condor sightings to help officials make management decisions. Other steps also were taken in 1965 to strengthen condor protection. The Condor Advisory Committee was assigned to advise the regional forester on all condor matters, an expansion of its responsibilities. In addition, representatives of the California Department of Fish and Game, U.S. Forest Service, U.S. Fish and Wildlife Service, and National Audubon Society undertook the first cooperative survey of the condor population. These people formed a committee that also included University of California zoology professor A. Starker Leopold and condor researcher Ian McMillan. McMillan brought special expertise to the proceedings, since he was one of the researchers who had determined that the bird was down to only forty individuals. The committee's goal was to conduct annual surveys of the condor population. The data would be used to determine trends in the condor population over several years.

In 1965, when the first alarum of a major condor decline was being sounded, the bird had no federal protection. However, conservationists such as McMillan hoped that a national law protecting vanishing species soon would be enacted. A concrete sign that it might be was passage of the first federal endangered species act in 1966. But although it authorized the secretary of the Interior to create a list of endangered species and to use federal funds for the purchase of habitat important to listed species, it provided no protection. The condor's only legal protection continued to come from the state of California, which brought fines of up to $1,000 against anyone convicted of shooting one of the giant vultures, along with sentences of up to a year in jail. Nevertheless, the federal departments of Agriculture and of the Interior did undertake some measures in behalf of the bird. In 1965 Interior assigned a full-time researcher to study wild condors. He was joined in his work in 1969 by a biologist from Agriculture's Forest Service. In 1970 the secretary of the Interior banned oil and gas leasing in the Sespe Condor Sanctuary, and the Bureau of Land Management put a moratorium on all mineral-leasing activities in areas important to condor survival.

Finally, three years later, condor management took a big step forward with passage of the Endangered Species Act of 1973. By 1975 the U.S. Fish and Wildlife Service had developed a condor recovery plan, and the survey committee had become the recovery team. Nevertheless, by 1978 the condor still numbered no more than forty. Ominously, the Audubon Society that year published a report by a panel of condor experts that declared it more likely that only twenty to thirty remained. In its report the panel also evaluated the federal recovery plan. It praised the Fish and Wildlife Service for seeking to put more condor habitat under government protection. It agreed with the call for better control of pesticide use within condor range, since contamination of the birds had been implicated as a possible cause of reproductive failure. But it criticized the plan for being short-sighted on a major point, the state's announced goal of striving to create a stable population of only fifty birds. The panel said that setting such a low goal "is to attempt to maintain the species precariously on the brink of extinction rather than to give the species a reasonable chance for self-perpetuation with minimum human assistance."

Although the goal was later revised to one hundred birds, the recovery plan soon led to scalding debate. The conflict stemmed from an emergency management proposal outlined in the recovery plan. Called a contingency plan, it was slated for use only if the recovery plan failed to arrest the condor decline. The contingency proposal troubled many conservationists because it emphasized the trapping and handling of condors. It even called for taking wild condors into captivity for a breeding program. Such intensive, hands-on work drew the ire of such groups as the Sierra Club and Friends of the Earth.

The nail finally was hit squarely on the head in 1979, when the federal Fish and Wildlife Service announced a plan that seemed to embody all the measures of the contingency proposal. The service said it intended to capture all surviving condors, radio tag and release some of them, and keep a selection of younger birds in captivity. The service was even practicing its capture and release techniques on Andean condors and on black vultures, a nonendangered U.S. species. Since the service seemed committed to captive breeding, many conservationists prepared for a fight. Legal action was threatened, and the Sierra Club passed a resolution opposing the plan. The debate waxed emotional, with some condor enthusiasts reacting as if the condors were holy entities and trapping them a sacrilege. Many feared that the birds were too fragile to be handled, that they might respond by vanishing abruptly into extinction. A number of biologists, Koford among them, had helped create this anxiety by suggesting that condors and their range be kept as isolated as possible from human intrusion, since intrusion would lead to nesting failures, abandonment of habitat and food, and a host of other problems.

If trapping and handling the birds seemed bad to some conservationists, radio tagging seemed positively evil. The radio tags in question weighed about two ounces and transmitted signals that biologists could pick up on special receivers. The tagging would allow researchers to track the birds as they moved around their territory. This would allow new insights into how condors use their habitat, where they roost and feed, how they interact, and a vast array of other activities. Or so argued those in favor of tagging. Opponents saw it as an invasion of the bird's privacy at best and as a threat to its life at worst. They argued that the transmitters could interfere harmfully with the bird's behavior.

But if radio tagging was evil, keeping the birds in captivity was positively unthinkable. The harshest opponents of captive breeding offered two basic arguments. The scientific argument was that taking condors into captivity would reduce the wild population, possibly pushing it to extinction. The emotional argument was that the condor was a creature of the skies, that it should dwell in freedom, not in zoos, that when soaring on outstretched wings it was a symbol of life's eternal struggle against doom, while in a cage it would be, as Koford wrote in 1953, uninspiring, pitiful, and ugly.

Those who supported all or a portion of the Fish and Wildlife Service plan took a different tack. They said the bird was dying off in the wild anyway. Taken into captivity, it might live and breed, as the condors of the South American Andes had done. If so, biologists might use the bought time to remedy whatever it was in the environment that was killing the birds. Then captive-reared condors could be released into their old haunts. If the entire project failed, the condor would be no more extinct than it would have been if left alone.

Opponents came up with an answer to this, too—death with dignity. They said it was better for the condor to vanish into the wild than to die huddled in a cage. This purely emotional argument seemed to suggest that the condors would simply live out normal lives, soaring over the wilderness and feeding on whale carcasses until they met a natural end. At the time, however, pesticides, shooting, lack of food, and loss of habitat to housing and development had all been suspected as causes of the condor decline. Given this situation, it is impossible to avoid asking how a condor that is killed by starvation, poisoning, or poaching can be said to have died with dignity.

Many of the emotional arguments against condor capture were of questionable merit. But the scientific reservations about capturing the birds had greater validity. Obviously, trapping condors for captive breeding would reduce the wild population. But if the birds vanished anyway, in hindsight it would have been regretted that captive breeding had not been attempted. On the other hand, if captured birds failed to breed, it would have suggested that the animals should have been left in the wild. Probably the truth was that the condor's existence had become so precarious that any steps taken on its behalf were bound to be tremendously risky. It was really a matter of which risk should be taken: capture, or forever free?

The National Audubon Society eventually joined the Fish and Wildlife Service in its bid to implement the contingency plan, pitting the society against such allies as the Sierra Club and Friends of the Earth. Audubon's role was important to the success of the contingency plan, since the society had been involved in condor research for decades. Long before the service had applied to the California Fish and Game Commission for permits to trap condors, Audubon had signed a cooperative condor management agreement with the Fish and Wildlife Service, Forest Service, Bureau of Land Management, and California Department of Fish and Game. Society biologists also had been working hand in hand with federal researchers at the Condor Research Center in Ventura, California. And Audubon had joined the American Ornithologists' Union in forming a special California Condor Scientific Review Committee to keep an eye on recovery work. With this behind it, Audubon's participation lent weight to the Fish and Wildlife Service plan. For the same reason, the Audubon presence must have seemed especially threatening to those opposed to the plan.

While the California Fish and Game Commission mulled over the decision to permit condor trapping, something new and positive occurred. It was, in effect, the lull before the storm. On May 14, 1980, biologists from the Condor Research Center hid themselves with brush while perched on a cliff in southern California. The biologists were intent on studying a condor nest high up on another cliff that was more than a quarter of a mile away on the opposite side of a canyon. The nest contained an egg laid March 15. From their concealed position, the biologists saw something no scientist had observed before, the hatching of a condor chick. Since they knew when the egg was laid, they were able to deduce for the first time the condor incubation period. This shows how little was known about the condor only a few years ago. The Department of the Interior lamented this lack of knowledge in a news release about the hatching, at the same time commenting pointedly on the need for condor trapping and radio tagging: "Years of conventional 'hands off' research have yielded little reliable data on the problems plaguing the condors because of their extensive range over rugged terrain and the great mobility of the birds."

The hands-off era was about to end, however. Fifteen days after the hatching, the California Fish and Game Commission made its momentous decision. Over the objections of various conservation groups, the commission approved the plan for trapping, radio telemetry, and captive breeding. At the same time, another federal project important to the overall contingency plan was moving ahead. This was the release into their native habitat of six Andean condors reared in captivity at a Fish and Wildlife Service research center. The Andean project was to serve as a prototype for California condor releases.

The crowning event in this promising period came on June 28, when researchers handled their first wild California condor, a nestling they called IC–9, for Immature Condor–9. The chick was vigorous and strong, and handling did not seem to affect it. Then the lull ended and the storm began.

It hit a month to the day after the capture permit was issued. A research assistant was handling and measuring a two-month-old, 13-pound condor chick while in the company of biologists from the Audubon Society and the Fish and Wildlife Service. Unlike the first chick, this one seemed to grow faint. Then, almost without warning, it collapsed and died. The death was a significant loss, since that single chick represented half the condors hatched that year. Subsequent examination suggested that it had died of shock.

The state of California immediately revoked its permits. A hearing was held the following August to decide the fate of the contingency plan. If permits were not reissued, the plan was dead. Friends of the Earth quickly stepped in to warn that the captures had advanced too quickly. They suggested that other species, such as the black vulture, should have been used more extensively as surrogates before condors were handled. By the end of July the Fish and Wildlife Service had decided not to reapply for a state permit until it had thoroughly studied the cause of the chick's death. It also stopped all research that involved the handling of condors.

While hands-on research stalled, other work continued. By April 1981, four captive-reared Andean condors were aloft in South American skies. This boded well for the California condor program. If Andean products of captivity could survive in the wild, perhaps captive-reared California condors could too. That summer, the Fish and Wildlife Service reapplied for state permits to trap condors. The state agreed. Under the new permit, which was good for three years, up to nine condors would be captured for breeding stock and a dozen more would be radio tagged. The Fish and Wildlife Service joined the Audubon Society in planning to trap California condors that fall.

The birds were to be baited with livestock carcasses. When they landed by the bait, researchers would trigger rockets to which a net had been attached. When the rockets blasted into the sky, the net would careen into the air and over the birds. Condor researchers had tested the nets on field trips to South Africa, where they had trapped vultures almost the size of the condor. The service planned initially to radio tag only two birds. If the birds survived without harm, more would be tagged. The service also hoped to net three birds for use in captive breeding. The Sierra Club, Friends of the Earth, and several local chapters of the National Audubon Society opposed the plan. The Sierra Club, however, gave lukewarm support to the use of radio tags.

Capture of the birds proved difficult. Nets were set in the fall of 1981 and spring of 1982, but they remained empty. Meanwhile, a development in the field opened a new avenue for acquiring captive stock. In February a pair of condors squabbled in their nest, breaking their egg. Loss of the egg was particularly tragic, since each condor pair hatches only one egg every other year. However, biologists were not completely without hope that this pair would breed that spring, since many bird species lay a second clutch of eggs if a first clutch is lost. Researchers watched the nest diligently and were rewarded April 7 by the laying of another egg.

Now that the biologists knew condors would double clutch they came up with a new scheme for boosting the birds' reproduction. They proposed to take eggs from the wild birds' nests and hatch and rear the young in captivity. Because the adult pairs presumably would lay and hatch another egg in the wild, the birth rate would be doubled. The Fish and Wildlife Service requested that its trapping permit be expanded to include the taking of eggs.

Meanwhile, in August 1982, the first wild condor, a nestling, was captured under the contingency plan. Researchers said that they took the young bird because it was neglected by its parents. It was sent to the San Diego Wild Animal Park for rearing. At the time of its capture, only one other condor was being kept in captivity. This was a male sent to the Los Angeles Zoo in 1967 after it was abandoned by its parents. In 1982 it was still alive and well. Part of the federal plan to capture breeding stock called for finding a female for it.

By the end of 1982 the netting program had paid off. Biologists had tagged and released two condors, released a third untagged, and kept one for captive breeding. The following January, captive-breeding plans received a boost when the California Fish and Game Commission expanded the federal government's permit to allow taking eggs. By the next summer, four eggs taken from wild nests had been hatched in incubators at the San Diego Zoo. There the chicks were raised with as little contact with people as possible. They were fed by keepers from behind a barrier so that the chicks could not see them. The keepers wore a lifelike puppet of an adult condor's head and neck on their hands and forearms, dropping food to the chicks from the puppet beak. It was hoped that this would help the chicks to learn to recognize their own kind, crucial for breeding. In most species, young animals learn this recognition from contact with their parents in a process biologists call "imprinting." If raised by humans, many animals become imprinted on people. As adults, they court and attempt to breed with humans and fail to recognize their own species. If the condor chicks raised in captivity became imprinted on people, the whole breeding project would be doomed.

By the end of 1983 the contingency plan seemed to be moving ahead smoothly. It was still being debated caustically in conservation circles, but nevertheless the captive population had grown to nine birds. It was even discovered that condors would triple clutch, producing a third egg if the first and second were removed. Then something went wrong in the wild population.

It began in March 1984, with the death of a five- or six-year-old male condor in a remote section of the Sierra Nevada. Then, between November 1984 and the following April, six condors died. Three were from breeding pairs, leaving only two breeding pairs where a year before there had been five. This was a massive and deadly blow to the wild population, thought at that point to number as few as eleven birds. The captive-breeding project also was dealt a harsh blow. Researchers had hoped to have collected thirteen eggs by early April but had only two. The condor recovery team began to ponder the need for another crucial decision: Should all remaining condors be taken into captivity?

The Fish and Wildlife Service decided in December 1986 that all surviving wild condors, thought to number about seven, would be brought in. At this point the National Audubon Society broke with the service, joining the Sierra Club and Friends of the Earth in opposing the capture. Audubon feared that the federal government would lose its incentive for protecting the birds' habitat if all the wild condors were captured. The society was particularly concerned because sensitive negotiations were under way to purchase the condor's main feeding ground and turn it into a federal sanctuary. The feeding site was on the privately owned Hudson Ranch. Two years before, Congress had appropriated nearly $9 million to buy the 14,000-acre ranch, but the deal had not been closed. Now the landowner was considering instead a plan to put a housing development on the ranch. If the condors were gone, the society believed the purchase would collapse.

The society also believed that wild birds were vital to the survival of any condors released from captivity. The society argued that released, captive-reared birds would need wild condors to guide them throughout their range, showing them where to find food, roost sites, and other needs. The society also thought that a reservoir of wild birds would provide insurance against something going awry in the breeding program. Already a potentially serious problem had occurred at the Los Angeles Zoo, where five captive condors had been intentionally tamed. This meant the birds quite likely could never be released into the wild. Zoo officials maintained that they thought taming the birds was all right because they were not scheduled for release. Audubon, however, maintained that the zoo had made a grievous error.

The society received a court injunction to stop the federal government from capturing the last condors. But eventually it became clear that the birds were not going to survive. The cause of their rapid decline was unknown, making it impossible to correct any environmental problems. Only this was certain: All known condor deaths in those last grim days were related to feeding. The birds had died after eating poisoned carcasses or had succumbed to lead poisoning, presumably after eating bullets contained in the carcasses of game animals lost by hunters. Researchers had sought to correct this problem by establishing a few feeding sites with the clean carcasses of domestic livestock. Although this effort was fairly successful, it became increasingly clear that the condor's habitat was no longer safe for the birds. In the end, Audubon dropped its objections.

The last wild condor was captured on Easter Sunday 1987. He was netted at 10:15 in the morning, Pacific Time, after he landed to feed on a fetal calf and a dead goat that had been put out for bait. As soon as the net fell over him, biologists from the U.S. Fish and Wildlife Service and the National Audubon Society closed in. They carefully removed him from the net. They knew him well. He was a seven-year-old male known as AC–9, for Adult Condor–9. His parents were the two birds that had squabbled and destroyed their egg, leading biologists to the discovery that condors double clutch. Formerly, he had been known as IC–9 and had been the first wild condor that biologists had touched. Now that they had him, their work with wild condors was done. Finished too was the link between condor and nature that had sustained the species for thousands and thousands of years. Now its existence, if it continued, was in man's hands.

As soon as he was captured, the last wild condor was whisked away to the San Diego Zoo for two weeks of quarantine. Then he joined thirteen other condors held at the San Diego Wild Animal Park. These birds, and thirteen more held at the Los Angeles Zoo, are the species' final toehold on existence.

The National Audubon Society, of all the conservation groups in the nation, likes to characterize itself as the voice of reason, as the group that takes the lead in approaching conservation problems rationally. Perhaps in the case of the condor, reason did not fail the Audubon Society. Research on condor habitat has continued. A special computer program has been created to help locate the best release sites for captive birds. The Hudson ranch has been purchased for the better protection of condors. And, most importantly, in 1988 a pair of condors in the San Diego Wild Animal Park hatched an egg. The hatchling was the first California condor to be conceived and brought forth in captivity. It has been progressing, condor biologists say, as smoothly as can possibly be expected. If other hatchings follow, and if more eggs can be double clutched, condors may be released into the wild as early as 1993.

Regardless of the success of the condor program, the massive effort undertaken to save the bird shows how difficult it is to save such a slow-breeding and specialized creature, regardless of its majesty and its appeal to the human imagination. This difficulty should serve as a warning wherever human endeavors encroach upon the natural world. The proper time to protect any living thing is before it is endangered, not after.

CHAPTER 9

The
Bowhead
Whale

In 1848 a lone whaling ship sailed through the Bering Strait and into the Chukchi Sea. On board were Yankee whalers, the best whalers the world had ever seen. Along with their European counterparts they had wiped out the vast numbers of bowhead whales to the east. Now they were in search of what they hoped would be an untouched multitude of bowheads, an animal once so plentiful that whalers called it simply the whale. They followed the summer recession of sea ice, a dangerous experiment for men in wooden ships under canvas sail, ice floes all around.

They found what they sought. At least twenty thousand bowheads plied the chill waters. Spring would find them in the central and western reaches of the Bering Sea. As the ice broke and the leads opened, the whales would move north across the Arctic Circle, traveling not 30 miles off the Alaskan shore. They would feed through the summer in the Beaufort Sea, north of Canada, then retreat to the Bering Sea as the brutal arctic winter rushed south and the ice closed over the sea again.

The new hunting ground was a boon to Yankee whaling. The whalers wintered in Honolulu and San Francisco with their wives and children. An entire social system developed, with wives and children sometimes joining the whaling expeditions. When ships

145

met in arctic waters, they would gather together and the wives would travel in row boats from ship to ship, socializing and exchanging gifts. Crews would throw parties. Doubtless the conversation dwelled mostly on whales and on the fear that soon the whales would disappear forever and for good.

Whalers favored the bowhead, as they favored its relative the right whale, because it was slow-moving and fairly easy to kill and would float when dead. Also, bowhead mothers were extremely solicitous of their offspring. If whalers harpooned a young whale, they could be certain of a shot at the mother when she came to the youngster's aid. The hardest part of the hunt was getting close enough to throw a harpoon by hand. Bowheads are shy and quick to retreat when they detect unusual sounds, such as oars pounding the water. The whalers had to be stealthy. Once struck, the whale might rush to swim under the ice, but it was more likely to dive. Whalers knew that the deeper and longer the dive, the more exhausted the whale would be when it surfaced. Then they could harpoon it again, or approach alongside and stab it repeatedly with a lance until finally it died in a sea red with its blood. A quick death might take two hours.

The whalers wanted two things from the dead bowhead. One was oil, a good lubricant and a major fuel for lighting. The other was baleen, or bone, as they called it. Baleen is what bowheads have instead of teeth. Made of a substance much like fingernails, it grows in long plates suspended from the roof of the whale's mouth. Each plate bears a fringed edge. Feeding whales take in great mouthfuls of water containing vast numbers of tiny sea creatures, such as shrimp and other crustaceans. Then with their tongues they force out the water, but the baleen holds back the sea creatures. The bowhead produces the longest baleen of any whale, with some plates measuring up to 15 feet, though 10 feet is more likely. In the nineteenth century baleen was used, among other things, for corset stays and to provide frames for hoop skirts.

In 1859 the first oil well was drilled near Titusville, Pennsylvania. That was the death knell of the whale-oil trade. But bone continued to be in demand for several more decades. In some cases, whole whale carcasses—some more than 50 feet long and weighing more than 40 tons—were left to sink into the sea once the baleen was removed.

In 1871 the Yankee whaling fleet lingered too long in the arctic, searching for shrinking numbers of bowheads. Much of the fleet was trapped by the ice. The whalers were able to escape by marching over a frozen sea to vessels that had not been trapped. When they returned the following spring the whalers found most of the fleet crushed by ice or burned by Eskimos. Out of thirty-nine ships, thirty-two went down. This loss was doubly draining because the insurance companies that covered the ships were owned by the ship owners. The incident nearly finished Yankee whaling. Disaster struck again in 1876, when ice once more trapped a major part of the fleet. In 1880, the whalers switched from exclusively sail-powered vessels to ships powered by steam engines and sail. Despite more losses to sea ice in 1888, the fleet wintered safely in the arctic in 1890. The steamships found new hunting grounds where they could make larger kills. The number of whaling vessels therefore increased. Then, in the first decade of the twentieth century, spring steel replaced baleen. The price of bone fell from $5 a pound to less than fifty cents. The Yanks stopped their hunt of the bowhead. It would have ended soon anyway. Whales were becoming harder and harder to find. The onslaught, it appeared, had nearly wiped them out. Scottish whalers continued the pursuit because they used whale oil in jute production, but in 1913 and 1914 they brought back no whales. The hunt was over, the whale nearly finished.

SAVING WHALES

What happened to the bowhead happened to virtually every other species of large whale. Right whales, humpbacks, and grays were all nearly hunted into extinction. Even the blue whale, perhaps the largest creature that ever lived, has been almost wiped out.

This debacle occurred because whaling was completely unregulated and tremendously efficient. Whales were essentially the property of all nations. What one nation did not kill, another would. So each tried to get the biggest share it could. Whaling started taking a heavy toll almost from the time that commercial whaling first was begun by the Basques in the fifteenth century. The Basques initially hunted the right whales that cruised past their shores. They quickly wiped them out and had to sail far out to sea for more. Whaling also wiped out coastal whale populations in the New World. By the first half of the nineteenth century, Yankee whalers sailed far into the Pacific to hunt sperm whales because species closer to home were too scarce to make the hunt profitable. Whalers hunting with hand-thrown harpoons from boats powered by oars were wiping out whales worldwide. And then whaling rapidly became more efficient.

In 1870 a Norwegian invented a harpoon with an explosive head. It blew up inside a struck whale, sending shrapnel tearing through the whale's body and speeding the kill. Within two years, blue, fin, and humpback whales had become rare in Norwegian waters. In the late 1800s newly invented harpoon guns increased the whaler's range and permitted him to shoot from the decks of the big ships. And then, in the twentieth century, as engines replaced sails, came the apogee of whaling technology, the coup de grâce for whales everywhere—the factory ship. These massive vessels were served by smaller catch boats that hunted and harpooned whales. Carcasses were towed back to the factory ship, usually anchored in a harbor or bay. The whales would be laid out beside the ship and completely processed. This eliminated the need to return to shore with raw materials. Whalers could now stay out longer and sail farther. In 1926 came a major breakthrough, the factory ship with a stern slip that allowed an entire whale to be brought up on to the deck for processing. Now whales could be rendered on the high seas, speeding the process dramatically. As the technology improved, whale carcasses were inflated with air and left floating until the factory ship came along and picked them up. Meanwhile, the catch boats continued the hunt. Because they were engine-powered, they could pursue whales usually too swift for nine-teenth-century whalers, such as the blue whale. Eventually, heli-copters were used to spot whales for the killer ships, and radio transmitters were attached to dead whales so that factory boats could home in on the radio signal and locate the carcasses. Between 1910 and 1967, more than 300,000 blue whales were killed. Whale populations are nearly impossible to census with accuracy, but it has been estimated that during those years the blue whale dropped from roughly a half-million animals to per-haps fourteen thousand. The magnitude of the kill is amazing. In 1975 alone, whalers killed seventeen thousand sperm whales, more than four times the number killed by the nineteenth-cen-tury Yankee fleet in its best year.

Whalers realized by the middle of this century that they were killing off the animals they needed to sustain their industry. To fend off the extinction of whales and thus of whalers, fourteen whaling nations met in Washington, D.C., in 1946. They signed a convention that proposed to end overhunting and, at the same time, tried to establish a cartel to control the production of whale oil. The latter was a measure designed to prevent oil gluts, which had plagued the whaling industry in the 1920s and 1930s. The convention also set up an organization to achieve its goals, the International Whaling Commission. Headquartered in Cam-bridge, the commission has met every year since 1949. Under the terms of the conventions, it could open and close the hunting of various whale species, limit the areas in which hunting could occur, and establish size limits and quotas on the number of whales caught. It could also specify the techniques to be used. For example, the commission outlawed the use of cold harpoons, those without explosive heads, because they killed slowly and painfully.

The commission was required to amend the convention each time it wished to implement some aspect of its powers. This slowed its effectiveness, since three-quarters of the commission members had to agree to an amendment for it to pass. Moreover, any member could object to the adoption of an amendment and thereafter ignore it. The commission thus had no means of enforcing its regulations. Whale conservation was further hampered because of the commission's domination by whaling interests. Though quotas were set, they did not reflect the biological needs of whales. They were based instead on a system that assigned values to various whale species. The basic unit of value was the blue-whale unit. Each nation was allotted a certain number of units. Under this system, a blue-whale unit equaled, for example, three sei whales or one and a quarter humpbacks. A whaler seeking to use up four blue-whale units could therefore kill a dozen seis or five humpbacks. The system was seriously flawed because it treated all whales as one species. The needs of individual species were ignored.

The commission was so ineffective that in 1970 the federal government listed eight types of whales, including the bowhead, as endangered species. The flimsy protections offered by the commission came under increasing scrutiny thereafter. This scrutiny was motivated at least in part by a growing public concern for whales. Researchers were learning fascinating new things about the animals. Some biologists began to suspect that whales were quite intelligent, which made the animals more interesting to the public. Intelligence is hard to measure, however, so debate about whale intellect still rumbles. Other more obvious aspects of behavior attracted equal attention, however. The songs of the humpback whale, for example, drew considerable public attention, especially after a recording of the songs was released early in the 1970s. The songs probably have something to do with breeding or territoriality, since they are produced only by males and only during the breeding season. Biologists believe the songs are quite unlike anything encountered in other species. Mating calls in most species never vary, but humpback songs change from year to year. Whales off Hawaii, for example, will all sing the same song one year. The following year, they all sing a variation of the song. How the song is changed in unison and why is unknown, but the mystery has brought new fans to the whale.

As the whales developed a constituency in the United States, they also developed support in Congress. The first sign of this support was the 1971 Pelly Amendment to the Fisherman's Protective Act of 1967. Although designed to protect fish populations, the law also helps whales by permitting the president to ban the import of fish products from any nation that interferes with the effectiveness of the International Whaling Commission.

The next major federal step in whale protection was the Marine Mammal Protection Act, passed in 1972. This law made it illegal for U.S. citizens to kill marine mammals or to import products made from them, though it permitted Native Americans to hunt the animals for subsistence if the hunted populations were not depleted. It also put the protection of marine mammals squarely under the aegis of the federal government, something many states still resent. The federal law was seen as necessary, however, to provide consistent protection for wide-ranging species such as polar bears, seals, sea lions, and walruses.

Although it adds some teeth to whaling-commission protections, the Pelly Amendment is implemented at the president's discretion. He does not have to impose an embargo on an offending nation if he does not want to. The 1979 Packwood–Magnuson Amendment to the Fishery Conservation and Management Act gave the commission an additional weapon that fires automatically. It *requires* the federal government to halve the fishing rights in U.S. waters of any nation that diminishes the effectiveness of the commission.

With increasing public concern for whales and increasing clout thanks to U.S. law, the International Whaling Commission assumed more importance in the effort to save whales. In 1982 it flexed its muscles with tremendous vigor by voting twenty-five to seven for a ban on all commercial whaling. The ban, slated to begin in 1986, was to last five years, after which whaling might start anew pending the outcome of studies of whale stocks. Japan, Norway, and Iceland have sought to buck the ban. In 1988 each announced plans to kill whales under a clause in commission regulations that allows taking whales for research. The commission's Scientific Committee, which evaluates research proposals, ruled that the research plans had serious flaws and withheld its approval. The United States then warned Norway that it might cut the nation's fishing rights in U.S. waters if Norwegians killed any of the thirty to thirty-five whales called for in the research proposal.

Iceland was not threatened with sanctions, however, because the government claimed that Iceland agreed to reduce its research kill to sixty-eight fin and ten sei whales. However, some conservationists believe that the government was swayed by threats to the status of a U.S. military base on Iceland. Japan, meanwhile, resubmitted its proposal after modifying it to meet the Scientific Committee's criticisms. The United States might cut the Japanese take of fish in U.S. waters pending the outcome of a new ruling by the Scientific Committee. In any event, it is apparent that the International Whaling Commission will continue to meet resistance to its regulations. Of course, controversy is nothing new to the commission. And one of the most controversial issues with which it has dealt is the right of Native Alaskans to hunt the endangered bowhead whale.

SUBSISTENCE HUNTING

Every spring and fall the Inupiat people of arctic Alaska ease their boats into the sea to hunt whales. For the Inupiat, the hunt means more than food. It is a ritual that dates back at least four thousand years. It permeates their culture. Without the hunt, say the Inupiat, their culture would cease to exist. Anthropologists who have studied them agree.

The Inupiat live in an extremely harsh environment in which, prior to European contact, scant food sources kept their survival teetering on a razor's edge. The scarcity of food resulted in a major cultural emphasis on sharing that has survived into the present. Even when young men make their first kill, they are expected to give away the entire animal to an elder. This reinforces ties between generations. Sharing creates a network of binding social ties not only within villages but between them, because coastal villages share whale meat with inland villages that lack access to the sea. Residents of inland villages also join in the whale hunts. Thus, hunting the bowhead provides both sustenance and a profound cultural tie between individuals and whole villages. For these reasons—at least in part—the United States has resisted annual efforts by the International Whaling Commission to stop bowhead hunting.

The hunts are conducted in spring and fall during bowhead migration. In spring, the Inupiat pursue the whales in hide-covered boats that they paddle 10 to 20 miles out to sea. In the old days, they used hand-thrown harpoons tipped with blades made from walrus tusks. The whales still are killed with hand-thrown harpoons, but usually the harpoon has an exploding head. Dispatching harpooned whales by use of shoulder guns that fire exploding projectiles is also on the rise. In another adaptation of new technology, motors may be used to tow dead whales back to shore for butchering. In fall, the hunters tend to use aluminum boats, but the hunt is much the same. Observers of the hunt believe that an average of one whale is killed and lost for each one landed.

Whaling crews usually are put together by older, more experienced men. Younger men generally lack the wherewithal to outfit a boat and crew. The hunt thus gives special status to its leaders. It is also the center of religious ceremony, since successful whalers sponsor a ritual called *nalukataq*. The status and ritual associated with whaling add even greater cultural importance to the hunt.

From the early 1900s until the 1970s the Inupiat killed a little more than a dozen bowheads each year. This loss did not seem to jeopardize the whales, so the International Whaling Commission never attempted to control the hunt. Then oil development came to the arctic. Suddenly, and for the first time, the Inupiat had money. Even young men could afford to earn social status by outfitting a hunt. In Point Barrow alone the number of crews jumped from twenty-five in 1971 to thirty-six five years later. Villages that had never hunted before started doing so. With increasing frequency, hunters started using motorized boats and harpoon guns. In 1976, the Inupiat killed forty-eight bowheads and harpooned another forty-three. Those harpooned got away but probably died of their wounds. The death toll leapt by nearly a factor of six. More whales were taken than the Inupiat could even use. The following spring they killed twenty-six and lost seventy-eight.

These seemed heavy losses for a slow-breeding species thought to number no more than two thousand individuals. The International Whaling Commission became alarmed. It concluded that the whales would be wiped out by the new subsistence onslaught. At its meeting that June, proposals to control Inupiat whaling led to a major battle. Several nations wanted the bowhead hunt stopped immediately. But though the U.S. delegate supported a proposed ten-year ban on commerical whaling, he abstained from voting on the bowhead issue. The United States sympathized with the Inupiats' ancient hunting tradition and did not want to enflame the Inupiats by opposing the hunt. The ban passed anyway, but without U.S. support it could not long survive. The following December the commission held a special meeting at which the ban was killed and a quota put on the number of whales that could be taken. The Inupiat would be permitted to land a dozen bowhead and to harpoon and lose eighteen, whichever came first. In 1979 the quota was raised to eighteen landed and twenty-six harpooned. At about this time the Inupiat formed the Alaska Eskimo Whaling Commission to look after their interests. The commission was given responsibility for dividing the quota among the villages. It was hoped that putting the Inupiat in charge of the quota system would encourage their cooperation.

The issue of subsistence hunting did not die with the birth of the quota system. Every year nations such as Australia, the Netherlands, New Zealand, and Panama fought at International Whaling Commission meetings for a complete ban. And each year the United States sought to up the quota. In 1980, when the Scientific Committee called for a ban, the United States delegate responded by asking for the same quota as the previous year, eighteen landed and twenty-six struck. The request was defeated, but the debate became deadlocked. In exasperation the U.S. delegate declared that if no quota were set, he would recommend upon his return home that the United States unilaterally set a quota of eighteen landed and twenty-six harpooned. The issue was not resolved until the last day of the meeting, when U.S. conservationists successfully proposed a block quota of forty-five whales landed and sixty-five struck over a three-year period. The limit of whales landed during any one year was set at seventeen. The three-year quota passed, though the United States abstained from voting because the Inupiat were not pleased. When the bowhead quota again came up for renewal, it was expanded to forty-three whales harpooned over a two-year period, regardless of the number landed. No more than twenty-seven were to be harpooned in a single year. The next quota climbed further by setting a limit of twenty-six strikes per year from 1985 through 1987 and allowing any strikes not used in one year to be carried forward to the next, with a maximum annual strike of thirty-two.

The U.S. position on subsistence whaling was of special concern to ban proponents. They feared that if bowheads were hunted their recovery might be slowed for a century or more or might even fail. They also suspected that the U.S. position could create a demand for subsistence hunting elsewhere. Their suspicions have proved correct. Japan has attempted to evade the ban on commercial whaling by arguing that whale hunting within its own waters is subsistence whaling. Lack of U.S. support for a ban on bowhead hunting has thus meant a potential weakening of the commercial ban.

Some proponents of the bowhead ban have argued that subsistence hunting is a thing of the past. The Inupiat, they point out, are no longer using strictly traditional means of killing whales. Use of explosive harpoons and harpoon guns suggests that the old ways are already lost. Since the bowhead may be declining, proponents suggest that loss of the hunt by Inupiat culture may therefore be less important than the potential loss of the bowhead for all time. Hunters of European descent changed their ways with the development of wildlife laws, and the Inupiat can change theirs, too, in the face of a changing world.

This is a tough, pragmatic argument. It overlooks the many adjustments the Inupiats have already made to protect whales. It also overlooks the commitment of the Inupiats to preserving the hunt. Anything as culturally important as the bowhead hunt will not be lost without a fight. The United States has fought Native Americans before with stunning success. But unlike the Natives of the Great Plains a hundred years ago, the Inupiat will not arm themselves with arrows and ride horseback into the face of rifle fire. They will instead fight back politically and strategically. For example, the Alaska Eskimo Whaling Commission, run by older, respected whalers, apparently has done a good, honest job of administering the quotas and setting high standards. But if the Inupiat were told that the hunt must end, they might simply ignore the edict and kill more whales than even the quota system allows. The government would be hard pressed to stop them, since it is difficult for outsiders to enforce wildlife laws in the Arctic. The federal government, in its dealings with Alaskan Natives, therefore has concluded that it is far easier to seek cooperation than to attempt to enforce unenforceable laws.

Aside from whale protection, the government is concerned about another factor. Arctic waters show some promise for off-shore oil exploration. Needless to say, oil-development advocates within the government do not want to arouse Inupiat anger over whales and have it turn into resistance to the presence of oil wells in arctic seas. Development could be slowed by, for example, lawsuits that seek to stop oil exploration on the grounds that it will harm the bowhead and other natural resources, such as waterfowl, polar bears, gray whales, beluga whales, seals, sea lions, and walruses. Conservationists have already argued that seismic exploration, in which loud sounds aimed at the ocean bottom are used to locate possible well sites, may frighten whales off feeding grounds. Conservationists also worry that ship and airplane traffic and permanent wells will force the whales away from presumably critical migratory routes and feeding sites. Oil spills are also a potential threat. Spilled oil could inflame the whale's hide, effect feeding efficiency by clogging baleen, poison whales through ingestion or inhalation, and kill off the plankton upon which whales feed. Persuasive scientific evidence suggests that these concerns are not serious, and a federal study indicates that the threats are minimal. But because some critics doubt these findings, the situation is volatile. By placating the Inupiat, the federal government may be hoping to avoid any possibility that the Natives will end up in the conservationist camp.

The bowhead hunt appears to be a permanent fixture in the Arctic. The oil companies have even integrated it into their development activities. Several of them recently signed an agreement promising the Inupiat that oil-company boats would avoid areas in which whaling was under way. This promise was made after the Inupiat became angry when intruding company vessels frightened off whales the Natives were stalking. Under the agreement, the Inupiat were given marine radios so that they could communicate with oil-company vessels. A few of the whaling boats were even given satellite navigation devices so that they could avoid areas where seismic boats were working. The oil companies also agreed to help whalers in emergencies. For example, when a harpoon head exploded inside a boat and blew off a whaler's hand, the oil-company communications equipment was used to call in a helicopter that helped get the injured man to medical aid. In another case, a helicopter was used to rescue whalers trapped at sea by a storm. The company ships may even help whalers tow in dead whales. However, they are not permitted to help hunters locate or kill whales.

The case of the bowhead whale shows how greatly values have changed in the United States. Barely a century ago, the federal government was running roughshod over both native peoples and wildlife. Now it is caught in a wrenching collision between the needs of an endangered species and the survival of an ancient culture. Times have changed, values have changed, but the problems created by a thoughtless past linger on. Consequently, even the good news about bowheads is dampened by bad. In 1987 a new technique for censusing bowheads indicated that the whales number about 7,800 animals, nearly double the previous highest estimate. But in June 1988, the International Whaling Commission boosted the quota on harpooned whales to forty-one. This is thought to be equal to about half the number of bowheads born each year.

CHAPTER 10

On the
Endangered
Species Act

It is difficult to imagine a more ambitious wildlife law than the federal Endangered Species Act. On the surface, it seems simple and direct. It declares that it is illegal for the federal government to fund any activities that harm listed species or reduce their chances of survival. This includes a myriad of activities from building dams that would destroy listed fish to draining swamps important to listed amphibians, from logging to plowing grasslands, from housing development to offshore oil drilling. It includes *any* activity that puts a listed species at risk. The law also prohibits private citizens from harassing, capturing, or killing all endangered and most threatened species. Because of these prohibitions, the Endangered Species Act has been called the most powerful wildlife protection law in the world.

The really amazing thing about the law is what it attempts to do when carried to its logical extreme. Virtually any species that has sunk almost to extinction can be nominated for listing. If the Fish and Wildlife Service, or, in the case of marine species, the National Marine Fisheries Service, finds that the candidate for listing is indeed on the brink, the animal or plant will be listed. Once listed, every effort is supposedly made to save the species— to recover it, as the bureaucrats say. The Endangered Species Act

159

has thus outlawed extinction, a biological process that has been going on for about two billion years.

This act truly is something new under the sun. It says several very interesting things about humankind. On one level, it says that people are extremely optimistic, particularly about the potency of their laws. They have sought to legislate away extinction. This is not a very pragmatic idea. Will we prevent the extinction of species for ten years, twenty, a hundred? For a millennium? If the idea begins to sound absurd on the scale of a mere thousand years, it becomes ludicrous when postulated for ten thousand (about twice the age of civilization) and impossible to discuss beyond that because it defies logic. Furthermore, we know that living things evolve. Genes mutate, plants and animals change. They become something else. Not all dinosaurs disappeared, for example. Some became birds, if we are to believe the fossil record. How does this stack up against extinction? Did those that survived in the guise of birds really become extinct? Or did they just become avian? If the Jurassic had had an Endangered Species Act, would zoos today have brontosauruses?

On another level, it is this complete lack of pragmatism that makes the act so appealing. Or if not the act, the people that support it. For the act shows that there are among us people who cannot look upon a suffering or dying thing, or a vanishing species, without a pang of regret and sorrow. They *are* something new under the sun. They are the first living things to care about creatures other than themselves. They have extended sympathy and compassion beyond their own immediate needs. They are so stricken at the loss of a species that they wish to stave off extinction. The Endangered Species Act thus shows that there is hope for human survival. If people such as these can ever dominate the actions of society, then perhaps the actions of society will cease to threaten the survival of all life.

It was intense optimism and a profound reverence for life that led to the passage of the Endangered Species Act of 1973. The law was worded so strongly that virtually any project could be stopped if it jeopardized a listed species. Such a sweeping concept alarmed the nation's business, agricultural, and development communities. These communities encompassed a great variety of people, including oilmen; loggers; ranchers; real estate developers; farmers; hunters; dam builders; the military; the Army Corps of Engineers; fishermen; producers of toxic wastes, pesticides, and air and water pollution; the federal Forest Service; the Bureau of Land Management; even the Department of the Interior, where the Fish and Wildlife Service is headquartered. Needless to say, few of these people are not known for their deep concern about wildlife and the environment. They certainly were not cheering in the galleries when the Endangered Species Act was passed. Instead, they feared that the act would scuttle some of their plans, costing them the civilized world's two most valued commodities: time and money. Their fears hardened their opposition to the act. And since many are politically and socially powerful, their concerns frequently have been transformed into congressional action that put new limits on the protection offered listed species under the act. Amendments that weakened the act also weakened the resolve of conservationists to use it too aggressively. For example, they feared invoking the act to stop the hunting of listed grizzlies in Montana because Congress might later respond by reducing the law's protection of all threatened species.

A perfect example of this—in fact, the classic example—occurred in 1975 in a confrontation between a tiny endangered fish and a dam on the Tennessee–Tombigbee River. The Tellico Dam was supposed to provide electricity, flood control, and recreation, with concomitant economic benefits, for eastern Tennessee. Thus spake the Tennessee Valley Authority, the federal agency in charge of the dam. The dam had been under construction for eight years at a cost of about $50 million. Then, in 1975, less than two years before the dam was scheduled for completion, the Fish and Wildlife Service added to the nascent endangered species list a fish called the snail darter. It lived in the river where the reservoir would be located. The service predicted that the dam would wipe out the darter and its habitat. The dam would have to be stopped, and private citizens sued to stop it. The legal battle raged all the way to the Supreme Court, which ruled in favor of the darter. Costs be damned, the court seemed to say. The Endangered Species Act demanded the protection of listed species.

The fight over the darter and Tellico Dam is still deeply regretted among conservationists. For one thing, opponents still bring it up as an example of the act's impracticality. Also, because the case was widely covered in the news media, Tellico Dam is the only thing many citizens really know about the Endangered Species Act. Years later, they still think that the act is merely a shield for obscure fish and an unnecessary encumbrance to progress. But worst of all, in the end the Tellico fight was brutally lost. Congressional action exempted the dam from the provisions of the act, so it was completed. Congress in 1978 also amended the act to slow the listing process and to cut back on the power of the Fish and Wildlife Service to protect the habitat of listed species. In addition, a special exemption process was created to help projects stalled by the act. The amendments also gave specific exemption for building a dam in Wyoming, whose construction was threatened by the presence of endangered whooping cranes.

This experience taught conservationists, to their sorrow, that the act could not be invoked too quickly, too loudly, or too powerfully. To protect the integrity of the act, conservationists had to use it gingerly. Ironically, the use of the act at Tellico had little to do with protecting an endangered species. The dam was an economic disaster, built at a great loss of federal dollars. Opponents of the dam wanted it stopped on purely economic grounds, but at the time Congress did not consider a high price tag sufficient reason to stop a dam. So dam opponents thought it would be clever to invoke the Endangered Species Act. It was really a ploy.

The act has suffered further reductions of its powers since 1978. It also faces major threats every few years when it must be reapproved by Congress. Passage of the latest version was held up for nearly two years by senators opposed to such provisions as the protection of listed sea turtles from shrimp fishermen, who drown thousands of the turtles in their nets each year.

This evidence shows plainly that the Endangered Species Act is only as strong as its most ardent and vocal supporters. If developers speak more loudly than conservationists, the act will falter. Similarly, it will flourish if elected officials are persuaded, by a vocal constituency, that wildlife protection is a heartfelt and critical concern of the voting public. Wolves would be roaming the distant corners of Yellowstone today if the outrage of disenchanted wildlife enthusiasts shook Congress as deeply as the outrage of wolf-hating ranchers. The act's greatest strengths do not come from Congress or the Fish and Wildlife Service or the latest presidential administration but from the will and words of those who actively support it. If the act fails, aren't they to blame?

Oddly enough, much of the fear that the act inspires in industry, business, and agriculture is for all practical purposes entirely unjustified. Only .25 percent of all the federal actions investigated by the Fish and Wildlife Service in 1984 were ruled likely to harm a listed species. A special committee created by Congress in 1978 to provide exemptions for projects deemed worthy of completion despite any harm they may do to listed species had had only one case brought to it by 1985. If anything, the act has not been enforced strictly enough. For example, because local citizens were opposed, critical habitat has never been designated for the grizzly bear, even though the law requires the Fish and Wildlife Service to designate critical habitat for all listed species. Nesting colonies of red-cockaded woodpeckers, a species endangered by loss of its southern forest habitat, have been intentionally destroyed on U.S. Marine bases with scarcely a whimper from the Fish and Wildlife Service. Sometimes implementation of the act has been intentionally stifled. The Reagan administration, for example, sought to cut funds for listing and for federal endangered species grants to the states. Listing has proceeded so slowly that it will take nearly a century for all current candidates to be investigated by the appropriate federal agency. By then, some will simply have dropped out of existence. Moreover, listing has tended to favor appealing creatures. Mammals and birds are more likely to be listed than insects and other invertebrates, even though invertebrates are vastly more numerous and no less ecologically important. The act is, quite clearly, a human product, marked by human foibles.

If the act has a glaring philosophical weakness it is this last point, that it focuses primarily on species useful and appealing to people. The drive to save living things would reach further and wax more promising if human society sought also to save things it does not like and cannot use and does not understand. If we could be as concerned about some shapeless lump of flesh writhing in the dust as we are about bowhead whales and condors, we could be much more confident of our ability to save wildlife and ourselves.

Perhaps one step in this direction has been the development of state nongame programs. Although nongame has been the focus of wildlife management agencies in some states for several decades, in the vast majority of cases nongame has been ignored. Even in states with nongame programs, funding has been scarce. This started to change in 1977 when the Colorado legislature added a check-off box to state income tax returns. Recipients of tax refunds could check the box if they wanted to donate part of their return to a new nongame management program. Taxpayers could also add a contribution to their bill.

The check-off was an instant success, raising $350,000 its first year—about three times the funding allocated by the legislature in previous years. The amount soared to nearly three-quarters of a million dollars by 1981. By then other states were jumping on the bandwagon. In less than ten years, nearly all states had created a nongame check-off program. Many traditional wildlife managers were disgruntled at having this new responsibility foisted upon then. Nevertheless, they had to respond. It was clear that citizens were not only interested in nongame management but were willing to pay for it voluntarily out of their own pockets.

Part Three

Nongame
Wildlife

CHAPTER 11

The Western
Diamondback
Rattlesnake

The early settlers wasted no feelings on rattlesnakes. They blamed rattlers not only for threatening human lives but also for killing livestock. Western ranchers organized "snake bees," competitive hunts in which people vied to kill the most rattlers. During a hunt in Iowa in 1849 two men killed ninety rattlers in an hour and a half. The entire hunt brought down nearly four thousand.

Nothing is known about how these hunts affected snake populations. A century ago or more, when settlers were doing their best to wipe out grizzlies and wolves, the rattler was beneath consideration. No one collected data on rattlesnake populations. If a survey were made of rattler numbers today, it would be impossible to know how the reptile's present populations compare with those of the pristine past. It is fairly certain, though, that rattler numbers have dropped in the vicinity of human settlements. The organized hunts probably reduced local populations on the ranches and farms where they were held, too.

No one knows how many rattlers survive in the West, or whether rattlesnake populations are decreasing, static, or increasing. This would be unimportant were rattlers not still trapped and killed mercilessly in organized hunts in various western states. Rattlers are one of the few species subject to virtually unrestrained and widespread commercial hunting with no data to show whether they can sustain the losses.

Whether rattlers survive or vanish altogether may not concern most people. Yet an interest in the fate of creatures such as the western diamondback rattlesnake separates those with a desire to protect the integrity of the natural environment from those interested simply in the preservation of appealing creatures. The diamondback is as important to its scrub wood and rangeland environment as are deer or ducks or native grasses. Rattlers are useful to people, too, since they kill millions of crop-damaging rodents each year. There is no denying that the rattler has an evil and sinister look. But that is a purely human interpretation with no roots in fact. Conservation should not be predicated on such irrational responses, and ignorance should not be permitted to determine which species survive and which fail. At the heart of modern conservation lies the conviction that all living things have a vital role to play. It is not necessary to identify that role before seeking to protect a beleaguered species.

THE SNAKE HUNTS

The only study on organized rattlesnake hunts ever released was conducted in Oklahoma in 1988. Funded by the state nongame program, it has provided the first publicly available data on the effects of the hunts. In addition, Henry Fitch and George Pisani, the two University of Kansas herpetologists who conducted the study, plan to publish several scientific papers on snake behavior based on the vast quantity of data the study allowed them to compile.

Rattlesnake hunts are held in spring, when the snakes are emerging from hibernation and are concentrated near den sites. Hundreds of western diamondback rattlers may congregate at a single den, making it easy for hunters to locate and collect large numbers. Some hunters pour gasoline into dens to drive out the snakes. This can permanently ruin a den site. Since rattlers tend to use the same winter dens year after year, loss of a den can create serious survival problems. The practice of gassing dens is therefore deplored by many collectors.

In recent years some Oklahoma snake enthusiasts, including trained herpetologists, have become concerned that the hunts are killing too many snakes and ruining too many dens. They have demanded that the Oklahoma Department of Wildlife Conservation, the agency that regulates hunting in the state, stop the hunts until data are collected to show that snake populations are not being demolished by overkilling and loss of dens. Knowing full well that it could not end the hunts without incurring a major social and political debt, the wildlife department responded in 1987 by funding the snake-hunt study. Decisions about the snake hunt would be made after the data came in.

Rattlesnake roundups, as the hunts are called, are held in five Oklahoma communities. Sponsors tout the hunts as educational, but critics call them carnivals. Posters advertising the roundup held in Waurika in the last week of March 1988 suggest that the critics are right: Among the main events of the two-day festival was a carnival featuring fifteen amusement park rides. A snake-sacking contest, butcher shop, and photo booth, among other events, shared the spotlight. Before the hunt began, four rattlers were marked with tags and released. They were worth $25 apiece to any hunter who retrieved them. Other prizes included $200 for the longest snake caught, $75 for the snake with the most rattles, trophies for the three shortest snakes, and a trophy for the longest caught by a first-time hunter. To qualify for the contest, all rattlers had to be brought in alive.

The roundups feature a variety of souvenir booths. Snakeskin wallets and hatbands are sold. So are Plexiglas toilet seats with whole rattlesnakes embedded in them. Fried rattlesnake is offered at food stands. The hunts always feature a pit crawling with dozens of rattlers. A snake handler periodically enters the pit, generally a large area enclosed by four-foot-high plywood, and strolls among the writhing reptiles. Since it is spring and the snakes have only recently emerged from hibernation, they are sluggish. The handler may kick them or prod them with sticks to get them to respond in typically snakelike fashion. Most do— which actually takes away some of the melodrama, since a snake's typical reaction to danger is flight. Once in a while, though, one will strike. But because diamondbacks tend to aim for a man's ankle, the handler is fairly safe. He wears boots.

The butcher shop, spelled *shoppe* on the Waurika poster, is a big draw for kids. For around a dollar anyone can step up to a butcher block and lop the head off a live rattler held down by a snake handler. Small children are not too adept at this, but that does not stop them from diligently chopping away at a snake's neck with a pocket knife. Dead snakes are skinned on the spot, with spectators watching as the mottled hides are pulled from the white-muscled bodies. The skins are sold at a set price per foot, and the meat goes to the food stands. Gall bladders may be shipped to the Far East, but more likely end up in California or New York, where they are much in demand among practitioners of traditional Oriental medicine.

Special exhibits highlight each roundup. One critic described these in the *Newsletter of the Oklahoma Herpetological Society*: "Before the animals are killed, they are usually used as part of an exhibition of cruelty and stupidity beyond belief. These snake-handling exhibitions are called educational by the handlers and sponsors. The only way they could be described as educational is by students of abnormal psychology." The exhibits include stacking coiled rattlesnakes on a handler's head to see how many can be added before the pile topples. Another popular exhibit involves putting several rattlesnakes into a sleeping bag with a handler. The handler is supposed to demonstrate the proper technique for crawling out without being bitten. The event itself is among the major causes of snakebite in Oklahoma. Another event that escalates the rate of snakebite is the sacking contest. A pile of rattlers is released on a flatbed truck, and contestants race to see who can bag up the most in the least amount of time. Some roundups give "White Fang Awards" to participants who get bitten. One seasoned handler from Texas claims to have been bitten more than forty times in the course of his career.

The roundups are often promoted as a way to control rattlers, so captured snakes are supposed to come from the surrounding countryside. They don't always. Some Oklahoma hunts cannot get enough snakes that way, so they buy them from other locales. Some snakes come from Texas roundups, which are held earlier than those in Oklahoma. Snakes also are sold from hunt to hunt within Oklahoma.

Several of the roundups offer spectators bus rides to nearby dens. There they can watch hunters catch snakes and can even try the activity themselves. This may not be as dangerous as it sounds, since diamondback rattlesnakes probably are much overrated as a hazard to people. Given a chance to do so, rattlers will flee at the approach of a human. If surprised or cornered, they may strike. Usually, however, they first warn of their intention by buzzing with their rattles. This helps to save the snake's life. If rattlers always struck at threatening animals much larger than themselves, such as horses, cows, and people, they quite likely would not long survive. They would be using venom needed to kill prey. And they would risk a fatal attack from the animal they bit. The rattle saves them all this trouble. It warns the potential victim of the snake's presence. Wise potential victims then run off.

Despite their venomous bites, rattlers are relatively easy prey. They are such easy prey that mere beginners can quickly learn to catch and bag them, for rattlers, like most venomous snakes, are among the most helpless of dangerous species. Any number of animals can press an attack more swiftly and harrowingly. Rattlers may strike with blurring speed, but they travel slowly. Even if most snakes did not try to retreat when approached or threatened, they still would be easy to avoid since most people can walk faster than a snake can crawl. But what makes a rattler particularly vulnerable is its limited weaponry. It has only its jaws to fight with. Once its head is under control, it is defenseless. Hunters quickly render it helpless by pinning down its neck with a forked stick or a pair of tongs. After that it is relatively easy to dump it into a burlap bag. The biggest hazard is the hunter's carrying a bag too close to his body. A rattler may bite through the burlap.

The interest in sponsoring snake hunts is purely financial. Most hunt sponsors are completely open about that. The Oklahoma roundups attract ten thousand to fifteen thousand spectators, a big draw for a small Oklahoma town. And the big draw means big money. The Waurika roundup grossed $27,000 in 1987. Revenue from roundups has paid for a rescue van, firefighting equipment, and the fireworks for the annual Fourth of July celebration. In Mangum, Oklahoma, Girl Scouts earn roughly $1,000 each year selling carnival tickets. One Mangum high school class ran a roundup food concession for four years and made enough money to pay for a senior class trip to Hawaii. In Texas, where the biggest roundups are held, towns may rake in hundreds of thousands of dollars.

The money comes partly from hunter registration fees. It costs $2 to take part in the Waurika hunt, for example. But the real money is in the sale of the snakes. The movie *Urban Cowboy* is blamed—or praised, depending on the viewpoint—for having created a craze for rattlesnake paraphernalia among wearers of western garb. Entire businesses have been set up to buy hides and other parts. Texas hunters have estimated that a half-million rattlers were sold in their state in 1987 at about $50 apiece. Snakeskin boots sell for hundreds of dollars a pair. Another expensive item is a rattler band for western hats. The top-dollar model features a whole rattler head mounted in front of the crown, mouth open and fangs erect. The skin circles the crown, and the rattles adorn the back.

Other rattler products also bring in money. Meat sells for as much as $10 a pound. Laboratories pay for live snakes. A hunter may get a dollar for a snake head that sells for up to $50 when converted into a paperweight. Rattle earrings sell for $10.

The number of diamondback rattlesnakes being sold at roundups in Texas has alarmed some of the state's herpetologists and amateur snake enthusiasts. In response to their fears, the

Texas Department of Parks and Wildlife sponsored a study of the effects of gasoline on burrowing animals in the early 1980s. The study indicated that gassing was deadly to insects, spiders, snails, frogs, salamanders, turtles, lizards, moles, shrews, prairie dogs, weasels, mice, and other burrowing animals as well as rattlers. Ironically, of the species studied the most resistant to gasoline was the western diamondback rattler. The Texas wildlife department has never released the study to the public. It has reportedly lost all copies of it.

Nearly a million rattlers were sold in Texas in 1986 and 1987. Most were caught by pouring gasoline into dens. Jonathan Campbell, the University of Texas herpetologist who conducted the research on den gassing, said that even veteran snake hunters are becoming alarmed. They claim that they always have sprayed such small amounts of gas that snakes reuse treated dens the same year. They worry that newcomers are using too much gas, apparently ruining dens for many years and thereby cutting into the annual take. If this is true, then the number of snakes that die when unable to find usable dens as winter approaches may be escalating the annual kill to catastrophic heights. Yet the Texas Department of Parks and Wildlife has chosen to ignore the problem, maintaining that it can take no action because it has no data to show that the snake population is declining. This claim represents not scientific caution but a lack of commitment to wildlife protection. Sound wildlife management requires that animals must not be hunted if no data can be generated to show that the hunt is not harming them. Campbell, who believes that the western diamondbacks of Texas are in decline, recommends that the snake be protected as a game species so that the hunts can be regulated. He also thinks that gassing should be outlawed.

FRESH DATA IN OKLAHOMA

Unfortunately, the Oklahoma study has limited applicability, since what is true for Oklahoma may not be true for other states.

Results from the study cannot be used as a key to the status of rattlers elsewhere, but nevertheless the study gives some indication of what is happening in the field.

Researcher George Pisani believes that about twelve thousand rattlers, the vast majority of them western diamondbacks, are killed each year during Oklahoma roundups. But because home movies taken over the past fifty years suggest that the number of snakes being caught has remained fairly stable, the kill does not appear to be harming the population. Still, roundup sponsors are becoming more cautious. Recognizing that the snakes are a major source of local revenue, they do not want to see the snakes wiped out. Consequently, some roundups are adopting protective measures. Some have done away with the shortest-snake contest category to prevent the taking of animals at a prebreeding age. The Waurika and Mangum roundups ban gasoline and firearms from hunt areas. Not all parties to the roundups have shown a similar enlightenment. A brochure published by the Shortgrass Rattlesnake Association, Inc., headquartered in Mangum, points out in a discussion on hunting techniques that pressurized spray cans with ten-foot nozzles are sometimes used to fire gasoline into snake dens, but says nothing about discouraging the practice.

The study recommends that state controls be put on the use of gasoline and that no more towns be permitted to start roundups. For the present, though, the hunts apparently are no threat to Oklahoma diamondbacks.

Snakes, particularly rattlers, are among the most vilified, condemned, defamed, maligned, and hated of all animals. This is a pretty strong reaction to an animal that generally weighs less than a Pomeranian lap dog, rarely exceeds five feet in length, and cannot see over a grown man's shoe unless it lifts up its head. But the vehemence with which snakes generally are viewed makes the Oklahoma study all the more important. It shows that conservation is moving beyond a concern only for things beautiful and appealing. However, the resistance to rattlesnake protection in Texas shows that a new era in conservation has not yet arrived.

CHAPTER 12

The
River
Otter

When the colonists first came to North America, they brought
with them a longstanding aversion to otters. In England the otter
had been treated for centuries as a pest. Apparently, it earned this
enmity by poaching from nets set for fish and by raiding the
ponds in which fish were kept until wanted for the table. Docu-
ments from the thirteenth century record the use of hounds to
track and kill river otters. By 1566, local governments offered
bounties on the animals. One town even required its fishermen to
keep a dog on hand for hunting otters. The fishermen were
required to hold at least two hunts yearly or pay a ten-shilling
fine.

In the New World no one seemed particularly distressed by
river otters, perhaps because fish were so bountiful. Or perhaps
the colonists found it difficult to worry about a three-foot-long
aquatic member of the weasel family when they had wolves,
mountain lions, and bears to deal with. This does not mean that
the river otter escaped unscathed. Early in the nineteenth century an otter hide could fetch $5 in the fur market, roughly the
same as a beaver or buffalo hide. It was outpriced only by the
grizzly's hide, which went for about $10, and the sea otter's hide,
which sold for up to $60. Clearly, the river otter's marine cousin

bore the fur of choice, so dense that when the animal was immersed in seawater its skin remained dry. But the river species also sported a dense and shiny fur that made a fine trim for royal robes.

Initially, trapping did not play a major role in the colonists' lives. Instead of trapping furbearers themselves, they bought hides from the Native Americans, who captured the animals by spearing, netting, or shooting them. The Natives also used deadfalls. A deadfall is a trap that is set by balancing a heavy object, such as one end of a log, on a small stick to which a bait is attached. When an animal tugs at the bait, it trips the stick and the heavy object comes crashing downward, crushing the animal's head.

The colonists brought with them a trap new to the Native Americans. This was the steel leghold trap. The trap had two springs affixed to a pair of jaws. Each spring was made from a long, flat piece of metal that was folded in two, like a pair of tongs. And like a pair of tongs, each spring lay open unless squeezed together. One end of each spring was attached to another strip of metal that formed the base of the trap. The other end was fashioned into an open ring. Each side of the trap jaws, which were hinged to the base, passed through one of these rings. When the springs were forced shut, just as a pair of tongs is squeezed shut, the jaws fell open. Then a small tab of metal, attached by a hinge to the base, was passed over one of the jaws and wedged under another hinged piece of metal called the trap pan. When an animal stepped on the pan, the tab was dislodged, the springs burst open, and the jaws snapped shut, clutching the animal's leg. The trap's design dates back at least to the 1500s, though various modifications have been made. A trap for a raccoon may have only one spring about 4 inches long. A bear trap would have two springs, each about 3 feet long, and would snap shut with enough force to break a man's leg. Traps for larger animals usually were armed with toothed jaws as well to give them better gripping power.

The traps were set both on land and in water. On land, traps were buried beneath a thin layer of soil. A scent, such as the urine of the animal sought, would be sprayed near the trap as an attractant. When an animal came to investigate, it would step into the trap. A trap set in water was usually placed where an aquatic creature such as a beaver or muskrat might crawl over it, such as at a den entrance. The preferred technique was to set the trap so that the captured animal would be held underwater and drowned, thus preventing damage caused to the hide if the animal struggled.

Depending on the strength of the trap, a captured animal's skin might be split or its leg broken. Some animals fight the traps, biting at them and breaking their teeth. River otters have earned a reputation for doing this. Trapped otters are often found dead from exhaustion. Many trappers believe that captive animals sometimes will gnaw off a foot to escape. It is not clear if this ever happens. Some observers think that an animal that thrashes violently when caught will break its leg and eventually rip off its foot, giving the impression that it chewed itself free.

The early laws of Massachusetts Bay Colony and New Plymouth Colony required all towns to make and set traps. This measure was taken primarily to catch and kill predators, particularly wolves. Use of steel traps to catch furbearers did not become common in the New World until settlers in Virginia and the Carolinas, perhaps more iconoclastic than other colonists, started competing with the Native Americans. Apparently, this broke a barrier, because furbearer trapping quickly became widespread. As otters, beavers, and other furbearers were wiped out in the East, trappers pushed west. Pursuit of furbearers thus proved instrumental to the exploration and settlement of North America. Trappers were the first Europeans to reach many parts of the continent. Before the eighteenth century drew to a close, trappers had reached at least as far as the Rockies. When Lewis and Clark arrived in North Dakota in 1805, they found French trappers waiting for them.

At first the trappers had problems with the traps they used. Most were made in England of poor-quality steel, causing springs to break when used in icy streams in winter. The first reliable, standardized trap made in the United States was the Newhouse trap. It was designed by Sewell Newhouse, who had started making traps when he was seventeen. In 1820 his family moved to Oneida County, New York. There he met the Oneida Indians, who were impressed with Newhouse's traps because they did not break in ice water. When the Oneida Indians were removed by the government to Green Bay, Wisconsin, they took their Newhouse traps with them. The traps quickly caught on in the West. After Newhouse joined the Utopian Oneida Community in New York in 1849, he started making a small number of traps each year. The demand for his traps was so great that in 1855 he opened a shop that employed three men. By the mid-1870s he had nearly a hundred employees and was turning out 300,000 traps a year. It was a lucrative business. Otter traps sold for at least $2 each. The income could quickly mount up, since most fur companies gave each of their trappers up to ten traps. Moreover, the value of a trap increased as it moved west, where traps were scarce. In the Far West, a trap could bring up to $20.

The new, reliable equipment increased the trapper's efficiency. Between 1821 and 1891, nearly a half-million river otters were killed for the fur trade in the United States. The Hudson's Bay Company took about 800,000 in the same period. The river otter thus receded throughout its range. It also suffered from wetlands loss and pollution. For example, the otter survived in fairly large numbers in Pennsylvania up to the late 1800s. After that, pollution from tanning factories and strip mines combined with trapping to reduce the population. Its numbers fell most precipitously in the midwestern prairies and the arid Southwest. The largest remnant populations occurred on the coasts and in the Great Lakes region. Today it is listed as endangered by eleven states. It is not on the federal endangered species list, but since 1979 it has been protected by the Convention on International Trade in Endangered Species of Wild Fauna and Flora. This is a treaty signed by more than ninety nations, including the United States, in an attempt to protect species whose survival might be harmed by trade. For example, trade in some very rare species, such as cheetahs, is completely banned, while trade in less vulnerable species is closely monitored. The river otter falls into the second category. Otter hides sold for export are supposed to be tagged so that the state of origin can be traced. The otter was given this protection because its numbers were thought to be low, and little data about its population levels existed. Under these conditions, it was suspected that the otter might be susceptible to over-hunting.

THE OTTER TODAY

Otters are members of the weasel family, a group that has a reputation for bloodthirstiness and predatory skill. Early naturalists rarely seemed to tire of telling how a tiny weasel could bring down a rabbit several times its size. But weasel or not, the otter never earned much stature as a predator. Instead, its reputation seems to be one of playfulness. It is probably best known as the creature that enjoys slipping on its belly down mud slides along river banks. It eats mostly fish. Bestselling books have been written about otters as pets and popular movies made from the books, making the otter probably the best known member of the weasel family. It seems to enjoy the easygoing, amiable aura of a porpoise with fur.

It seems quite odd, then, that very little is known about the otter's biology or about what it needs to survive. The staples of its diet have been only scantly studied. No one knows how many river otters live in America today, whether the population is increasing or decreasing, or how they are affected by human activities. Despite the lack of data, several factors are recognized as key to the otter's survival.

As a creature of streams and marshes, the otter is dependent upon wetlands. Its future is therefore intimately tied to the decline of wetlands nationwide. With wetlands disappearing at the rate of nearly a half-million acres a year, the otter's future must be considered anything but assured. The channelization of rivers is also harmful to otters, since they prefer streams that are slow-moving and that have ample cover—shrubs, trees, and other vegetation—along the banks. Channelization ruins both by straightening the bends that slow river currents and by cutting down bank vegetation. To all intents and purposes, a channelized river has been turned into a ditch. The water runs swiftly through it, and most deep pools are eliminated. The absence of streamside vegetation allows runoff from rain storms to pour swiftly into the stream. The runoff carries heavy loads of silt, which increases the stream's turbidity and allows it to capture more heat from sunlight. Channelization can also wipe out fish that require clear, cool, slow-moving water and can reduce the number of birds that nest along river banks. The effect on otters is not well known. Presumably, the otter population declines when wetlands and natural streams are lost. Studies that show how otters use various types of wetlands and streams, that map the areas in which otters occur, and that show which wetlands and streams should be saved for otters, whether currently used by the animals or not, are desperately needed.

One development that has helped the otter is the return of the beaver. Like the otter, the beaver was wiped out in the eighteenth and nineteenth centuries over much of its range. Once found throughout the United States, it was trapped into rarity to supply materials for the hat trade. Protection of remnant beaver populations and relocation of beavers from areas where they existed in stable numbers to areas where they had been destroyed helped the animal rebound. Today it is found in secure numbers in most states.

Beavers, like people, manufacture their own habitats. By building a dam on a stream, a colony of beavers creates a pond. The pond gives the beavers a place to build their dome-shaped houses and raise their young. The pond also has far-reaching effects on the beavers' environment. It permits the growth of aquatic plants, such as water lilies, upon which beavers feed. It also drowns any trees growing along the edges of the dammed stream. Eventually, the pond fills with silt or the dam breaks and the beavers move on. If the dam was built in a wooded area, the drowning of the trees and the draining of the pond create an opening in the woods called a beaver meadow. Such meadows are beneficial to a variety of wildlife. The beaver is thus useful to the survival of a wide variety of species. Even deer benefit from the forest openings. River otters benefit, too, from the ponds themselves. The beavers create the wetland habitat that otters need and that people have destroyed in so many areas. Yet the relationship between beavers and river otters has not been well studied. Research into this relationship could yield both a better understanding of the river otter's needs and a clearer picture of the beaver's role in the natural environment.

Studies also are needed on how pollution affects river otters. The otters' fish diet makes them very susceptible to pollutants because fish feed on insects and other small creatures that become loaded with pesticides, heavy metals, and other toxins. The pollutants build up in the fish and are passed on to the otters. The effect of pollutants on otters, as well as the levels of toxicity that otters can survive, is unknown. Some biologists believe that if current trends in pollution continue, the otter will probably decline.

Trapping is another problem that besets the otter. Until the otter was protected under the Convention on International Trade, the state wildlife agencies lacked reliable figures for determining whether otters were being overtrapped. This was a serious concern in the late 1970s and early 1980s, when the price of otter pelts rose to $60 apiece. When prices are high, trappers make a greater effort to take a particular species. If trapped species are not monitored, they can quickly disappear during times of inflationary prices. When the federal government decided to add the otter to the list of species protected under the international trade agreement, several states were spurred to undertake studies of otter populations. However, it will take several years of collecting data to determine population trends.

Being trapped for fur is not the only problem otters face. Otters in the United States have followed the example set by their British cousins centuries ago: They sometimes raid fish hatcheries. Some states allow hatchery personnel to kill otters year round to protect their fish. Another cause of mortality is nontarget trapping.

Trapping is not a very selective way to capture wild animals. For example, a man hired by the federal government to trap coyotes in Oklahoma captured two skunks, an opossum, and a dog in the process of bagging three coyotes in a day of trapping. Otters fall victim to traps set for beavers, which are commonly sought by trappers and share habitat with otters. For this reason, trapping can account for the loss of river otters even in areas where they are protected.

Some sport fishermen are also a plague upon the otter. They see the otter as a competitor for limited prey. In some states, fishermen have even demanded that wildlife agencies remove otters from streams where the anglers believed the animals were taking too many fish. Evidence suggests, however, that anglers' fears are groundless. A study made nearly fifty years ago showed that although otters can swim fast enough to overtake a healthy trout—one of the faster game fish—they feed primarily on slow species such as suckers, minnows, and other fish uninteresting to anglers. They also feed heavily on amphibians. Game fish form up to a quarter of the diet or more in some areas, but usually the otters focus on small individuals that an angler would not want. Moreover, a 1955 study indicated that otters feed primarily on whatever fish species is most abundant, making it unlikely that they will take too many individuals from any given fish population. Results from this study suggested, too, that otters may benefit sport fishermen by removing fish that compete with trout for food and space. More recent studies also have found that otters are no threat to game fish. Nevertheless, the otter continues to draw the enmity of some fishermen. Like hunters, the anglers believe that their license fees paid for the fish, so the fish should be all theirs.

Aside from this segment of the angling community, otters are popular animals. Their apparent good nature and their cuteness have won them a lot of allies. This has made them the subject of a number of projects funded under nongame programs and designed to reintroduce otters into areas from which they have been exterminated.

REINTRODUCTIONS

Many different types of animals, from turkeys to deer, have been returned to emptied habitat by reintroduction. These projects require an amazing amount of bureaucratic cooperation, since most wildlife agencies are attempting to reintroduce species that either are no longer found within their state boundaries or are not present in numbers great enough to provide individuals for relocation. Wildlife officials therefore have to go to other states for help. For example, early in the 1980s Missouri wanted to release otters into its streams. Most of the state's otters had been wiped out years before. But what Missouri lacked in otters it made up for in wild turkeys. So it worked out a complicated deal. It sent thirty-two of its turkeys to Kentucky, where turkey numbers were low. Kentucky then paid private trappers in Louisiana $8,000 for twenty river otters, ten of each sex. When the otters were secured, Kentucky shipped them to Missouri.

In addition to Missouri, otters have been reintroduced to suitable habitat in Arizona, Colorado, Iowa, Kansas, Kentucky, Minnesota, Oklahoma, Pennsylvania, and Tennessee. Most initial reintroductions were successful enough to stimulate state wildlife officials to begin additional releases. Reintroductions also are being considered in Illinois, Indiana, Nebraska, Ohio, and West Virginia.

One thoroughly documented reintroduction took place in the Great Smoky Mountains National Park, where otters had not been seen since 1936. They had been wiped out by trapping. Wildlife graduate student Jane Griess described the reintroduction in a master's thesis prepared for the University of Tennessee. Her description shows the tremendous amount of cooperation that typifies most relocation work, whether for otters, other nongame species, endangered species, or game animals.

The park's otter reintroduction required professional input or funding from several sources, including the National Park Service, Tennessee Wildlife Resources Agency, Tennessee Valley Authority, National Audubon Society, Tennessee Valley Sportsmans

Club, Great Smoky Mountains Conservation Association, and a number of private individuals. The project originated in 1981, when a study indicated that the park had about 100 miles of streams suitable for otters. In 1984, representatives of the park service, Tennessee wildlife department, Tennessee Valley Authority, and University of Tennessee met to discuss river otter reintroductions.

The project got under way in 1986, when the Tennessee Valley Authority hired trappers in North Carolina to capture ten otters. The trappers used steel leghold traps, the same sort of trap that had been used to wipe out otters throughout most of their range a century and more before. The traps had to be checked at least every twenty-four hours to help limit injury to the captive animals. The trappers performed better than expected. They caught sixteen.

Two otters died when the vehicle they were in bogged down in a snowstorm while on the way to a holding facility in Knoxville, Tennessee. The rest were transported by plane and arrived without mishap. Upon arrival in Knoxville they were examined by a veterinarian. All the otters had some type of injury as a result of trapping. The injuries ranged from minor cuts to severe damage, such as compound fractures that required surgery or other extensive medical treatment.

Because otters frequently die from stress when handled or moved, the captured animals were held in the holding facility for a week before being tagged with radio transmitters. The transmitters, surgically implanted in the otters' abdomens, were miniature versions of the radios with which the Craigheads had tagged grizzlies. The transmitted signals could be picked up by researchers on the ground or in airplanes, allowing biologists to monitor the otters' dispersal, food habits, and social life. Such knowledge was crucial to evaluating the success of the program, particularly since otters are subject to high mortality in relocation projects. Eighty percent of the otters released in one Arizona project died within two weeks. Four out of ten otters released in Oklahoma died within five weeks.

Eleven otters were released in the park between late February and the end of March. Within two weeks one otter died. An older animal with broken canine teeth, he apparently died of starvation. His death was the project's only bad news in its first year. A year after the first releases a male otter was caught in a fishing net and died. He was turned over to the University of Tennessee, where an examination proved that he had been in excellent condition. He had even gained weight, a sure sign of success for an animal that had weathered relocation from the warm coastal environment of North Carolina to the cold mountain streams of Tennessee.

Jane Griess's research shed some light on the nature of the otters' diet and may answer the concerns of some anglers. She collected seventy-five scats during her study and found the remains of fish in 90 percent of them. However, only two species of game fish were taken, rainbow trout and rock bass. The trout were found in 14 percent of the scats, the rock bass in scarcely more than 2 percent. Earlier studies had indicated that otters feed upon the most abundant fish available. However, the released otters apparently selected their prey according to speed, not abundance. The rainbow trout was the most abundant species in the stream in which the otters had been released, but 81 percent of the fish remains in their scats came from slower-swimming fish that made up a smaller proportion of the fish population. Most of the research on the otters' diet was conducted during spring and summer, when crayfish and amphibians made up a large share of the otters' food. In winter, when these animals are less available, exploitation of fish might increase and the proportions of species consumed might change.

WHAT LIES AHEAD?

Reintroductions are an important tool for increasing the number and density of otter populations. Acquiring a thorough knowledge of the otter's population dynamics and biological needs is also critical to the animal's survival, since the information is fundamental to establishing sound regulations for trapping. But clearly the most important factor in otter conservation is protection of wetlands from destruction and pollution. Without wetlands, there is no place in which to trap otters, and nowhere to release them.

As state nongame biologists compile more data on river otters, it is becoming certain that in some areas otters are increasing. Not only are some local populations growing larger, but otters in some states are creating new populations by returning on their own to streams from which they were extirpated long ago. Moreover, otter management seems to be at the dawn of a new day as growing numbers of studies begin to answer the many questions important to otter conservation. This development should put otter management on a reliable biological footing for the first time. But wetlands losses and increasing levels of pollution cast a grim shadow over the otter's future. The outcome need not be bad, however. Pollution is as threatening to human populations in some areas as it is to otters, since people, too, rely on streams for water. Eventually, people must respond to the threat as much for their own health and survival as for that of any wildlife species. This will benefit both otters and people. As biologist Paul J. Polechla, Jr., asserts in the *Audubon Wildlife Report 1988/1989*: "Although the long-term prognosis [for the otter] is bleak, negative trends can be reversed if we realize that properly managed, unpolluted watersheds are ideal not only for otters, but are also in the best interest of humans."

CHAPTER 13

Bats

Bats seem like biological oddities, since no one expects mammals to fly. But the numbers suggest otherwise. Some 4,000 species of mammal roam the Earth today. About 950 of them are bats. This means that nearly one in every four species of mammal has wings.

Only thirty-nine species of bats occur in the United States. A few feed on plant nectar, but most on insects. They are nocturnal, appearing with the dusk and vanishing with the dawn. Most are so small that they can be cupped in an adult's hand. And yet they are tremendously reviled and feared, victims of superstitious dread, of an ignorance that has bred contempt.

Perhaps the most damaging myth about bats concerns rabies. It is a common belief that most bats carry the disease. It is even widely accepted that a visitor to a bat cave can contract rabies by inhaling the dust from the bat dung that cakes the floor. Biologists have learned, however, that these age-old gems of folk wisdom are wrong. Only about .5 percent of the apparently healthy bats examined by researchers in the United States were infected with rabies. Only nine people in the United States are known to have contracted rabies from bats in the past thirty years. Only one case has occurred in Canada. Only one known case of rabies transmission from dung inhalation exists on record.

Even if bats frequently harbored the rabies virus, they still would pose little threat to people. They are active at night, when presumably they would have little contact with most of the human populace. They are shy. They are also extremely skilled at avoiding objects that might harm them, such as brick walls, telephone lines, and people. They are without doubt less likely to cause the average person harm than is the dog next door or, for that matter, the dog's owner. If people simply avoid touching any bats they find, the bats will do them no harm.

Bats are useful to people and important to the natural environment. They are a living insecticide. One large bat colony in Texas wipes out about 125 tons of insects every night. A single, small, insectivorous bat can gulp down a thousand insects in one night. A bat that lives for thirty years, as individuals of some species do, might account for over six million insects in the course of its life, presuming it takes off about five months of each year to hibernate for the winter. A larger bat would eat more.

Bats, like some birds, are critical pollinators of many plants. The agaves, plants of the desert Southwest used in making fiber ropes and tequila, are bat pollinated. One study found that in the absence of bats, seed production in some agave species crashes to only three-thousandths of the normal level. Both the giant saguaro cactus and the organ pipe cactus are thought to be dependent on bats for pollination, since their flowers are reproductively active only at night. Declines in desert bat species are thought to be a major cause of declines in these cacti.

Bats have been used in a wide range of medical research projects. They have been instrumental in the development of birth-control techniques, low-temperature surgery, vaccines, and navigational aids for the blind. But because so little is known about them, it is likely that their importance to the natural environment and to human society has scarcely been plumbed. Under current conditions many bat species may be gone long before science comes to a better understanding of them and of the value they hold for other living things.

BATS ON THE WANE

Certain biological characteristics make bats highly susceptible to sudden drops in population. They reproduce slowly. In many species, females do not begin to breed until two years old and then have only one young per year. Nature usually matches a species' reproductive rate with its level of mortality. Rabbits and mice breed rapidly because they die fast and young. The bats' slower reproductive rate works well for an animal with low mortality, which apparently is what bats enjoy when human activity does not press them too hard. But once an animal with the bat's reproductive strategy withers into decline, it recovers only with difficulty, if at all.

The vulnerability of many bat species in the United States is compounded because the animals gather in caves for at least part of each year. Few caves seem to meet the exact needs of the various cave-dwelling bat species, so when bats find a cavern they like they crowd into it. This makes it easy for large numbers of bats to be killed in one swoop. Often misguided or merely foolish people wipe out whole bat colonies by setting fires in their caves. Burning rubber tires also have been used to smoke out bats.

Several species of bat seem to be fading away in the United States. Perhaps foremost among them is the Indiana bat, a tiny animal that weighs scarcely a third of an ounce. Though it ranges over a large portion of the East in spring and summer, about 85 percent of its population winters in only seven caves. The cave ceilings may be packed with as many as three hundred hibernating bats per square foot. The bats prefer caves with a winter temperature of 36 to 46 degrees. Few caves meet their specifications. Missouri, for example, has more than forty-seven hundred known caves. Only twenty-four have ever housed more than a hundred hibernating Indiana bats at a time.

In 1973 the Indiana bat became the first of its kind to be placed on the federal endangered species list. It has been declining for at least a quarter of a century. In the 1960s, for example, Kentucky had five caves in which more than 30,000 Indiana bats hibernated, for a state total of 330,000 bats. In 1988 the total winter population for all five caves was only 49,000. One Kentucky cave censused in 1987 housed 250 bats, while in the early 1960s it had held at least 100,000. The cause of their disappearance was a gift shop that had been built over one entrance to the cave and an improperly designed gate that had been built over a second. The construction warmed the cave, ruining it for the bats. When the bats returned to it for winter hibernation, they could not use it. They perched instead on the walls of the gift shop, where thousands were scraped off and rolled away in wheelbarrows. The Indiana bat's numbers probably have been cut in half since 1980.

Human disturbance of caves used for hibernation, usually called hibernicula by bat biologists, is the most serious cause of the Indiana bat's decline. A single visit by people to a cave in winter can arouse hibernating bats, raising their metabolism to levels that can consume enough energy to sustain up to a month of hibernation. With an energy debt of this magnitude imposed upon them, the bats are at serious risk. Vandalism also has helped restrict the bats' choices for winter shelter. The problem is particularly deadly because bats tend to return to the same roost each year. When a cave is ruined, the whole colony that used it may vanish.

Federal efforts to protect the Indiana bat call for categorizing caves into several types. The most critical are Priority One caves, which hold more than thirty thousand bats. Only eight such caves are known. Priority Two, also important, hold more than a thousand but fewer than thirty thousand bats. Priority One caves are being protected with fences and gates that do not alter air flow or impede bat movement. Many of the Priority Two Caves also are protected. Nevertheless, bat numbers in the Priority One caves dropped 55 percent from 1980 to 1988, and Priority Two caves are faring no better. The cause is unknown. Perhaps it lies beyond caves alone. In summer the bats roost in hollow trees and under loose bark. They also give birth at these warm-weather sites. Some land-use practices call for cutting down dead, hollow trees, an activity that is affecting populations of cavity-nesting birds. Perhaps it is harming bats, too. Dam projects that flood river bottoms presumably demolish woodlands useful to bats, and stream channelization and agricultural development have also been implicated in bat declines.

Purposeful killing of whole bat colonies has helped constrict the animals' numbers. A recent example occurred in January 1988, when shotguns were used to destroy some one thousand bats in a Kentucky cave. The bats were an aggregate of several species. When the shooting was over, only five Indiana bats and a few bats of other species remained.

Another jeopardized species is the gray bat, once the most numerous mammal of the Southeast. Hundreds of thousands poured nightly from caves in Tennessee, Alabama, and Mississippi. During the Civil War, the South took bat guano from virtually every large bat cave it had, using the guano to make gunpowder after imports of saltpeter were cut off. Mining the guano helped the South prolong its struggle and caused heavy, if untallied, losses among the gray bats.

At the end of the war, the bats were given back their domain and they recovered. Then, about a century later, cave exploration started increasing in popularity, and bat colonies began to vanish. Some colonies disappeared within a year after being disturbed. The trend continued. Between 1970 and 1976 gray bat populations fell an average of 54 percent in twenty-two colonies surveyed by bat expert Merlin Tuttle, founder in 1982 of Bat Conservation International and a leader in bat protection worldwide. Although individuals in several colonies showed dangerously high levels of insecticides, PCBs, and lead in their tissues, it was clear that human intrusion was the main cause of the declines. Tuttle successfully sought federal listing of the gray bat as endangered in 1976. The Fish and Wildlife Service subsequently placed several caves under protection. Private groups, such as Bat Conservation International, and several state nongame programs helped with surveys and with the acquisition of caves. Cave protection is critical to the gray bat, since fewer than 5 percent of caves surveyed in the Southeast are suitable to the animal.

The gray bat shows some signs of recovery. Four caves used by bats for breeding in summer are protected. In 1976, bats were no longer breeding in three of the four. The number of bats using one of the caves had sunk from half a million, including breeding females, in 1969 to only 128,000 nonbreeding bachelors in 1976. Now breeding is occurring in all the protected caves, and bat numbers for all four are up to nearly three-quarters of a million.

Perhaps the most drastic declines of any U.S. bat species are occurring in the Mexican free-tailed bat. Colonies of this animal form what may well be the densest gatherings of any mammal, with roosts averaging about 1,800 adults per square yard. Free-tailed bat young, which do not roost with their mothers, crowd together at an even more astounding density: 5,000 per square yard. One cave in central Texas houses an estimated 20 million Mexican free-tailed bats. The animals migrate between the southwestern United States and Mexico. During the breeding season in the United States, the total population probably strains toward 150 million.

And yet their numbers have been dropping. Carlsbad Caverns, famous for its bat flights on summer evenings, has lost the bulk of its residents. Nearly nine million are believed to have summered in Carlsbad in the 1930s. Since the mid–1970s the number probably has averaged only half a million. In Arizona's Eagle Creek Cave, free-tailed bats dropped from twenty-five million in 1963 to only thirty thousand six years later. This last example shows clearly that large numbers do not ensure a population's survival when habitat is extremely restricted.

The cause of the declines is obscure. One factor may be insecticides such as DDT. DDT was banned in the United States in 1972, but nearly a million pounds are used yearly in Mexico. Residues of DDT have been discovered in Mexican free-tailed bat guano and in bat tissues. In some bats, pesticide levels in body fat theoretically were high enough to kill the animals when they used the fat during migration. The evidence is still too slim to pin any part of the blame for Mexican free-tailed bat declines on pesticides, but it is likely that the poisons are playing some role.

Human disturbance probably is the major factor in the various bat colony declines in the Southwest. At Carlsbad, for example, a shaft for a guano mine was drilled into the ceiling of the biggest roost. The shaft changed the roost's temperature, humidity, and airflow, evidently with deleterious effects on the bats.

Some Mexican free-tailed bat colonies may be recovering. The shaft at Carlsbad has been plugged, and bats are returning to their roost. Their numbers have not yet increased, however. Eagle Creek Cave had few or no breeding females in 1985, but a year later a substantial number of females and young were observed there.

SAVING BATS

A number of measures are being taken to protect bats. Perhaps one of the more effective is exclusion of people from bat caves. Properly designed gates have been built over the entrances to several major roosting caves, including Hubbards Cave in central Tennessee. One of the largest bat hibernicula in the world, this cave attracts migrating bats from at least six states. More than 75 percent of the nation's gray bats winter there. So do seven other bat species, including the Indiana bat. But Hubbards Cave has a long history of disturbance. It was one of the caves mined for guano during the Civil War, and in recent decades had been subject to frequent human visitation. Bat Conservation International has joined with the Tennessee Nature Conservancy, Tennessee National Guard, Mid-State Steel Corporation, Nashville Grotto of the National Speleological Society, and the Cave Conservation Institute to build a huge steel and concrete gate over its entrance. The gate is designed to exclude humans but to allow bats to pass through. It stands more than 30 feet tall and 35 feet wide and weighs about 130 tons.

Bats were never a major part of the traditional wildlife manager's business agenda, but the listing of several bats as endangered species has spurred interest in their conservation, and the state nongame programs have helped provide the funding for study and protection of the animals. Pennsylvania, for example, has put a gate over a cave used by Indiana bats and has spent three years surveying bat hibernicula. The state also has received federal funding to survey abandoned mines, which are increasingly important refuges for bats as caves are lost. Florida has surveyed all its known bat caves and has posted most of them to help stop human intrusion. It has also completed a gray bat management plan and has acquired the animals' most important cave. Both states have been distributing educational materials about bats. Missouri has for several years been distributing instructions for building bat houses, which resemble roofs without buildings under them erected on four tall wooden legs. More than a dozen other states have entered the arena, helping to promote a better understanding of bats through magazine articles about the animals and through the distribution of literature provided by Bat Conservation International.

The various programs undertaken by the states may help dispel the ignorance surrounding bats. This could be vitally important to bat survival, since ignorance yields not only bliss, but also fear and intolerance. The more that is known about bats, the more they are likely to be valued. The measures being taken to protect bat caves are another sign that mindless human actions will not be permitted to send bats foolishly into that endless night, extinction.

CHAPTER 14

Migratory Birds

Ruby-throated hummingbirds, great blue herons, chimney swifts, robins, cedar waxwings, meadowlarks, and red-winged blackbirds—in the United States they are thought of as North American birds. They are common throughout summer in U.S. marshes, woodlands, or plains. But if judged by where they spend the bulk of their lives they are not North American birds. They migrate each autumn to areas south of the Tropic of Cancer. There they spend a half to two-thirds of their lives. To a Latin American they are Latin American birds. They go north only to breed.

Astonishingly enough, 332 bird species that breed in North America migrate in and out of Central and South America. This number includes 20 species of duck and goose, 10 species of hawk, 4 species of falcon, 27 species of sandpiper, 13 hummingbird species, 50 species of wood warblers, 13 oriole and blackbird species, and 26 species of finches, among others.

These include the birds that biologist Rachel Carson warned about in 1962 when she wrote *Silent Spring*. In that classic of conservation literature, she argued that rampant use of pesticides to control urban and agricultural insect pests was poisoning the nation's birds. If use of the chemicals continued, she said, American springs and summers soon would fall silent, bereft of bird song. The pesticide industry scoffed at first and finally attacked.

They attempted to ravage her arguments. But soon other developments spoke more loudly than even her words. Bald eagle populations nosedived. The peregrine falcon declined all over the United States and vanished east of the Mississippi River. The osprey—a large hawklike bird of coastlines, streams, and lakes that feeds on fish—nearly faded away. Research showed that the one thing all these birds had in common was a high level of pesticides in their tissues. They had absorbed the pesticides from the food they ate—fish, rodents, small birds—which got them from insects and from plants sprayed with the pesticides. Further study showed that the pesticides were interfering with the birds' ability to produce eggshells. The eggs they laid had abnormally thin shells that broke easily. The weight of parent birds during incubation was enough to destroy them. In addition, some birds failed to lay any eggs while others laid eggs containing dead embryos. The pesticide most clearly implicated in the reproductive failures and the decline of the bald eagle, peregrine falcon, and osprey was DDT. In 1972 the federal government banned its sale throughout the nation. Since that time all three species have begun to recover. Doubtless, less studied bird species also benefited from the DDT ban. But for many of these birds the ban may have been only a stay of execution, not a reprieve. Research coming out of the tropics indicates that future springs may yet be silent.

THE TROPICS

In the United States, migratory birds are protected by such federal laws as the Lacey Act, which prohibits interstate transport of illegally taken wildlife, and the Migratory Bird Protection Act, which makes it illegal to kill birds out of season. Migratory birds also are the subject of a great deal of study in the United States. Many biologists work each year on research designed to provide information on hawks, shorebirds, songbirds, and other migrants. Since 1966, the federal government has attempted annual surveys of breeding populations of about five hundred bird species. The surveys do not provide an estimate of bird numbers but rather reveal trends in population increases or decreases and in range expansion and shrinkage. Birds are also protected by government regulations, such as the ban on DDT. For example, in 1988 the Environmental Protection Agency prohibited the use of the insecticide diazinon on golf courses and sod farms because it was killing large numbers of birds.

Any complacency that conservationists and bird enthusiasts felt because of these protections has been dispelled in recent years by the growing realization that the future of migratory birds is as closely linked with what happens to them in Latin America as it is with what happens to them in the United States and Canada.

For the past decade biologists have been coming to grips with the question of how migrant birds that breed in North America fit into the ecology of Central and South America. For many years it was assumed that the birds merely wintered there, attracted by balmy weather and abundant food. But biologists have begun to surmise that survival for migrant birds is greatly dependent on what happens to them in the tropics.

Researchers in Central and South America have discovered that when birds migrate from the north into the tropics they immediately assume ecological roles similar to those of birds that live there year round. Their behavior is indistinguishable from that of the full-time residents. For example, some migrant species set up territories in the tropics. A territory, in scientific terminology, refers to an area that an animal defends from intrusion by other members of its species. That migrants establish winter territories was news to ornithologists, who had presumed migrants were territorial only in nesting areas. It is a sign that food resources are limited enough to make their defense beneficial to individual birds. Territoriality in the tropics during nonbreeding seasons underscores the importance of these areas to the survival of the birds.

Even species that do not establish territories are faithful to the areas over which they range. Individual members of these species may spend their entire winter within a single locale. They may also return to the same locale each year. If these birds were to lose their feeding ranges they would be left without a place to live. They would become what one researcher, ornithologist John Rappole, calls floaters. These are birds that wander from place to place because they lack a home of their own. They tend to die off more quickly than individuals living within a defined range or territory. Individual birds with an established range presumably survive better because they know where to find food and shelter.

Migrant birds apparently have a distinct role to play in the tropical environment. Research has shown that some trees in Panama and Costa Rica fruit during times of year when migrants are passing through. The birds eat the fruits and their seeds. When they pass the seeds in their droppings, they help disperse the trees. During thousands of years of evolution, this association with birds apparently has made some tropical plant species so dependent on certain migrants for seed dispersal that they fruit only during peak migration periods. For example, the pokeweed around Veracruz, Mexico, fruits when Swainson's thrushes arrive. These thrushes are among the few species that eat pokeweed fruit.

The orchard oriole, which breeds in the United States, plays an important role in the reproduction of a type of Panamanian tree. It is the tree's best known pollinator. Several migrants are critical pollinators of a Mexican tree species. Similarly, when the Tennessee warbler feeds on the nectar found in the flowers of a tropical vine it aids in the plant's pollination.

Pollination and seed dispersal by migrants show how intricately the birds fit into the natural world of the tropics. This fit seems surprising only because northern biologists have long presumed that migrant birds evolved in North America and made long flights to the tropics only as an adaptation to a winter scarcity of food. New evidence, however, suggests that the birds evolved in Latin America. Seventy-eight percent of the migrant species have close relatives that breed exclusively in the tropics. Nearly half the migrant species have subpopulations that breed in the tropics. Both these factors suggest that the migrants originated in the tropics and later dispersed into the north, if only seasonally.

Why they dispersed is subject to speculation. Rappole believes both competition between individuals of the same species and predation by other animals is intense in the tropics. This makes the establishment of safe territories difficult for young birds. Consequently, some species—those capable of using a wide range of foods and habitats—expanded their ranges into the north, where reduced competition and predation permitted more successful reproduction. Presumably, the birds that did this were floaters that made a success out of range expansion. The pressures that drove the birds north are still at work, but the outcome of such expansion is not always good. Every year tropical birds are found in temperate zones far out of their normal ranges. Many of these birds die because they are unable to cope with the stresses of life in the north. The migrants that nest in the United States and Canada represent species whose tropical forebears carried enough adaptive flexibility in their genes to survive in a vastly different environment. But when winter comes, the migrants still seek their ancestral home, led by signposts still undiscovered by science.

What is happening in their ancestral home, and how it is affecting the birds' survival, is the subject of intense debate in ornithological circles.

THE DESTRUCTION OF
THE TROPICAL RAIN FORESTS

Less than a century ago, tropical forests may have covered as much as 210 million acres in Latin America. Although about two-fifths have been cut down, they still house between one million and six million species of plants and animals, concentrating vast numbers of plant species in relatively small areas. The United States has about twenty thousand species of plants, while Colombia, only a ninth as large, has 25 percent more plant species. Most grow in the rain forests, which cover only a third of the nation. In a quarter acre of Colombia's Choco rain forest grow some two hundred species of trees, while the entire state of

New Hampshire breeds only twenty-five. A single park in Costa Rica houses more bird species than all of North America. A single reserve in Peru supports more than five hundred bird species, scarcely three hundred short of the total number for the United States and Canada.

This multitude of species is a hidden treasure for humankind. For example, curare, used as an anaesthetic during surgery, comes from a tropical plant. Native Americans who live in Amazon forests use more than thirteen hundred plants for medicinal purposes. Scientists are studying these plants to see if they offer cures for disease. Mexican yams have yielded chemicals used in birth control and against such diseases as arthritis, sciatica, and dermititis. A fourth of all pharmaceuticals used in the United States come from tropical forests worldwide, including chemicals used in the treatment of heart disease and cancer. Thousands of plant species remain to be studied for possible additions to the pharmaceutical arsenal.

Because of its vastness, the Latin American rain forest is critical on a grand scale to human survival. It plays a major role in the natural cycling of moisture and in absorbing sunlight. It also is important to the cycling of carbon dioxide. The trees lock up millions of tons of carbon in their cells. When the trees are cut and burned for agricultural clearing, the carbon is released into the atmosphere. There it helps trap the sun's heat on the Earth. If the forests are cut and burned rapidly, releasing huge amounts of carbon, the result could be a general global warming that will lead to widespread disruptions in crop growth. The sea level will rise as glaciers melt, inundating coastal cities.

Currently, the rain forests of Central and South America are being cut at an alarming rate. In Central America land is being rapidly cleared by the agricultural and logging industries. By the turn of the century, less than 10 percent of Central American forests will remain if current trends continue. In some areas, such as coastal Brazil, 95 percent of the forests already have been cut. In the Amazon basin, one of the world's largest remaining rain forests, trees are being cleared at the rate of about 4,000 square miles yearly for housing as well as agriculture. Dams built on

Amazonian rivers are also wiping out huge tracts of forest. A single dam proposed for the Tocantins River is expected to drown nearly 800 square miles of forest. As many as thirty-five more dams are planned for the region. If the rain forests of the Amazon basin eventually are reduced only to areas now protected, more than two-thirds of the bird species in the area will be lost.

Research indicates that widespread deforestation of the tropics will cause major declines in up to a hundred bird species that nest in North America. This expectation is based on studies funded in part by the U.S. Fish and Wildlife Service in an effort to collect data on nongame birds. As forests are cut and migrants lose their winter territories and feeding ranges, more birds will be turned into floaters, with the attendant reduced survival rates. One study found that after a Mexican rain forest was cut, local bird populations dropped by nearly 40 percent. Such reductions are perhaps already being echoed by declines in North American populations of thrushes, kingbirds, warblers, tanagers, and vireos. A U.S. Fish and Wildlife Service analysis of bird survey data collected since 1966 indicates a decline in fifteen of sixteen bird species that winter in Central American forests.

Not all biologists ascribe to the idea that declines in migrants are caused by loss of tropical forests. Some argue that the birds are dying off because of fragmentation of U.S. forests into tracts too small to provide good nesting sites. Rappole and others counter with several studies suggesting that tropical deforestation is the real culprit. For example, if the declines in migrant bird numbers were caused by fragmentation of U.S. forests, then presumably similar declines also would be observed among species that remain in the northern forests year round. But research shows that year-round residents are not failing. Proponents of the forest-fragmentation theory counter that migrants are declining because their nesting behavior makes them more vulnerable to fragmentation than do the nesting habits of year-round residents.

One study that seems to support the idea that tropical forest destruction is killing off migrant birds was conducted by Eugene S. Morton, curator of birds at the National Zoo in Washington, D.C. He banded Kentucky warblers near Front Royal, Virginia,

in the foothills of the Blue Ridge Mountains. His banding research found that only 25 percent of the birds that arrived at
Front Royal each spring had been banded. Under normal conditions the return rate should have been closer to 60 percent. A
Kentucky warbler study on the Maryland shore corroborated
Morton's results: In the first year of study only one of eight
breeding males returned. Research at both sites took place in the
bird's best breeding habitat. Older, more experienced birds
should have been dominant there since younger birds tend to be
shunted off to less desirable sites. The low return rates of older
individuals suggest that the birds are declining because the rain
forests they need to survive the winter are vanishing.

Two surveys of breeding birds conducted yearly since 1948 by
the Audubon Naturalist Society in Washington, D.C., also suggest that something is amiss among migrant songbirds. In the
1960s the surveys started turning up sharp declines in several
nesting species. Predictably, the first to go were the hooded and
Kentucky warbler, American redstart, and Acadian flycatcher.
These species are all dependent in winter on the shrinking old-
growth rain forests of the tropics. But they were not the only bird
populations to show signs of failing. The surveys also discovered
drops in nine other species, among them the ovenbird, yellow-
throated warbler, red-eyed vireo, and wood thrush. A third survey, conducted by the society yearly since 1959, came up with
much the same results.

All three surveys were conducted in the Washington area.
Because the study area was restricted, the cause of the decline
could be purely a local problem. However, studies elsewhere also
have turned up songbird declines. At various sites in New York
State, a dozen species have declined. In Wisconsin, warblers,
orioles, tanagers, and vireos are among the missing. In a virgin
forest tract in West Virginia's Cheat Mountains, a census of
breeding birds conducted regularly since 1947 started turning up
declines in migrant songbirds late in the 1960s.

The histories of three other songbird species also suggest that
tropical deforestation may be at the heart of migrant declines.
One of these is the Kirtland's warbler. Its breeding range in the

jack pines of central Michigan was not discovered until 1903. By then it was rare even though suitable jack pine habitat has always been abundant in the northern United States and southern Canada. The warbler's limited range despite adequate habitat has been blamed on the cowbird, which lays its eggs in the warbler's nest. When the cowbirds hatch, they kill the Kirtland's eggs or hatchlings and are raised by the warbler. The theory that the Kirtland's has been reduced to only two hundred pairs by the cowbird is based on the idea that the cowbird was not found in the jack pines until agriculture helped bring them there. Its nesting activities are thought to be an environmental disruption with which the warbler cannot cope. Studies show, however, that the cowbirds reduce the Kirtland's breeding success by little more than what is found in other warbler species and cannot even account for annual fluctuations observed in the Kirtland's populations.

Loss of habitat in the tropics may have been the real cause of the warbler's decline. Its winter range apparently is restricted to the Bahamas. But settlement has subjected the area to extensive clearing for farming. By the middle of the nineteenth century, the forests of the Bahamas had been almost completely destroyed. Several species found only there have vanished. The same habitat destruction quite likely reduced the Kirtland's to its present precarious condition.

A similar case involved the Bachman's warbler, a species that was rare when discovered in the nineteenth century even though its summer habitat could be found throughout the southeastern United States. In seeking an explanation for the bird's consistently low numbers, ornithologists have turned to its winter range, the forests of Cuba. Such a restricted range put limits on the number of birds that could survive the winter, creating a bottleneck to population growth. When settlement came to Cuba the rain forests were converted to sugarcane fields. This destruction of 85 percent of the forests may explain why the Bachman's warbler is perhaps the rarest migrant songbird that visits North America.

Further evidence of the role played by tropical forest in migrant bird survival comes from the Swainson's warbler. Discovered in 1849, the bird was then so rare that it was not seen a second time for fifty years. Yet in the past few years it has become increasingly common in the swamps and thickets of the Southeast. The reason for its population increase may lie in the bird's diversity of winter habitat. Most Swainson's warblers winter in the lowlands of the Greater Antilles. But some also migrate to lowland Mexico and Belize. There, in recent decades, all but the most productive plantations have been abandoned. They are reverting to the forests from which they were carved. As new forest growth springs up, the Swainson's warbler presumably finds more habitat in which to live, and its numbers increase.

Many of the conclusions being reached about the role that tropical deforestation is playing in migrant bird declines are speculative. No one can point with utter certainty to the clearing of the forests as *the* cause behind dwindling songbird numbers. The situation is far too complicated, too fraught with variables. Loss of habitat has also been rampant in the bird's breeding range in the United States and Canada. By the end of the nineteenth century not 5 percent of the forests east of the Mississippi River remained. Bottomland hardwood forests have been flooded by dams and sawn into oblivion. In many southern states less than 25 percent of the bottomlands remain, areas that presumably could be critical to such birds as the Bachman's warbler. The plains, too, have seen massive disruptions as native grasses were overturned and buried by the plow. Today, many native grasses survive only in narrow strips along highways and railroad lines or in cemeteries. Presumably, the birds dependent on them have dwindled.

It would seem to the untrained eye that settlement of the New World has been a case of burning the habitat of migrant birds at both ends. Hearing ornithologists arguing among themselves as to whether tropical deforestation or loss of North American habitat is cutting into bird numbers yields an almost reflexive wish that each side was a little less sure of its own position and had a bit fewer doubts about the opposition's. Isn't it likely that *all* the figures compiled by breeding-bird surveys before the 1960s, when declines were first recorded, would themselves have appeared quite low if they could have been compared with figures compiled a century or two before? Isn't it therefore likely that the tropical deforestation of recent decades is merely delivering a *coup de grâce* to birds already reduced by loss of breeding range? Isn't it more than likely that healthy bird populations depend on intact range in both north and south?

Many of the doubts about the cause of migrant bird declines might have been dispelled had broad scientific surveys of breeding birds been made as little as half a century ago. This would have provided biologists today with the data needed to determine migratory bird population trends. But fifty years ago the federal agencies best equipped to deal with such surveys were interested primarily in counting ducks and geese. Consequently, ornithologists have no long-term, reliable figures against which to compare today's nongame-bird populations.

Fortunately, that situation has been changing. In 1965 the U.S. Fish and Wildlife Service initiated its annual breeding-bird surveys. They are conducted by biologists who count the number of birds seen or heard on a single spring morning as the biologists drive along some eighteen hundred randomly selected roads scattered across the United States. The number of birds recorded for each species is used to show trends or fluctuations in bird numbers from year to year. However, the surveys have serious flaws. Determination of population size for any species is impossible at least partly because the surveys are conducted primarily in areas that can be easily reached. Areas tending to bar access by motor vehicles, such as marshes and swamps, are ignored. Moreover, adequate data on populations is collected for only about half the five hundred species surveyed. Nevertheless, the surveys are a beginning in the collection of data on nongame birds.

In the past decade, the U.S. Fish and Wildlife Service also has funded a limited amount of research on the effects of tropical deforestation on migrant birds. One result of this funding was *Nearctic Avian Migrants in the Neotropics*, the report in which

John Rappole, Eugene Morton, and others argue that tropical deforestation is a primary cause of migrant declines. Much of this chapter was drawn from that work, which still stands as the most comprehensive source on migrant birds and deforestation. In 1983 the service also funded a study on the relationship between forest size and bird abundance in the tropics. Whether more funding will be forthcoming is uncertain.

State research is also benefiting migrant birds. Breeding-bird surveys are among the most popular projects funded by the new state nongame programs because birds have a large constituency. Each year some fifty million people go birding, as birdwatching is called by the true initiates of the sport. The states hope to build strong support for nongame programs by appealing to these people. Also, because so many people go birding, the states have at hand a large force of volunteers from which to draw when seeking to survey birds. Future generations therefore may not have to suffer from the lack of reliable data that has plagued ornithologists whenever they try to discern trends in nongame-bird populations.

The apparent trend already discovered by ornithologists may be a sign that bigger problems are on the way. Among wildlife, the first species to decline are generally those most sensitive to environmental changes. The Kirtland's and Bachman's warblers apparently are failing because their restricted winter ranges make them highly susceptible to changes in tropical habitat. They were thus the earliest warning that something was wrong in the environment. The present dwindling away of even less sensitive species suggests that the environmental problems are becoming more widespread. Eventually, they will reach to all creatures, including humans.

If the tropical forests are permitted to vanish in a final massive plundering of the Earth's natural wealth, humanity will lose more than economically valuable species and potential cures for deadly diseases. It also will lose the huge living machine that helps power the planet's water, carbon, and heat cycles. Climatic changes on a vast and deadly scale may soon follow, leading to far-reaching disruptions of human society. Future generations, living in a world so convulsed, may look back at the present in anger. A century from now, they may read of today's declines in migratory birds and see in that history an unheeded warning of global catastrophe, a missed chance to have avoided the ruin that surrounds them.

CHAPTER 15

Nongame Programs

I was working for a state wildlife agency the year the agency finally received approval from the legislature for a nongame checkoff on state income tax forms. It had taken about five years to get approval for the checkoff. A coterie of us in the wildlife department were elated that it had finally happened. As one of the first states to join the checkoff system, we were now in the vanguard of a new movement in wildlife management. For those of us tired of state biologists who thought wildlife conservation was game management and nothing more, passage of the checkoff was a boost to morale. It promised to provide a launching pad for new, far-reaching programs. Now no species would be ignored. The wildlife department could no longer ignore the plight of declining songbirds while it went on counting deer and setting hunting seasons.

At that time, nongame programs were the newest trend among state wildlife departments nationwide. Colorado's succcess with the tax checkoff sent interest in nongame programs soaring. Within a few years after the Colorado innovation, nearly every state had initiated some sort of nongame program. Not all were funded with tax checkoffs. Florida, for example, raised $2 million in less than a year for its nongame program by levying a special tax on the registration of cars previously registered in another state. But for the majority of states, the checkoff was the way to go.

Initially, the checkoffs met with outrageous success. For most states, hundreds of thousands of dollars rolled in the first year. In subsequent years, the figures tended to climb. It seemed too good to be true. And of course, it was.

A number of difficulties hinder the success of the nongame programs. One is a lack of commitment by many traditional wildlife managers to anything that has to do with nongame. From their viewpoint, nongame is not quite savory, not quite proper for a manager to be dabbling with, not quite, you might even say, manly. Lack of enthusiasm for the program from the people hired to administer it undermines the program's effectiveness. It keeps the programs from accomplishing as much as they otherwise might. One measure of this shortcoming is a comment, made by a state wildlife department public relations man in 1980, that his agency had raised so much money through the checkoff that the department did not know what to do with the cash. It was buying video equipment that no one knew how to use and funding studies of ground squirrels. This comment suggests that wildlife managers, trained to think in terms of game, simply do not know what to do with funds for nongame wildlife. Presumably, the agency would have had no problem figuring out what to do with a half-million fresh dollars given to it for waterfowl, quail, deer, or other game management.

Lack of commitment to nongame conservation and faulty training probably are not insurmountable problems. As time passes, traditional managers may be replaced by more innovative wildlife biologists interested in whole ecosystems, not just sport for hunters. The problems may simply vanish as these better trained managers arrive. But another problem is more persistent. It concerns the fundamental wisdom of funding nongame programs by tax checkoffs.

The checkoffs seemed like a dazzling idea just a few years ago. Giving people the chance to pay voluntarily for wildlife management would show game biologists how much the general public cared about wildlife. A fresh breeze might blow across the rather stagnant atmosphere of wildlife management. Unfortunately, it has not completely worked out that way.

The biggest snag is that the checkoffs do not provide a stable source of funds. The cash flow is subject to taxpayer whim. One year the checkoff in a given state might draw three-quarters of a million dollars, the next year half a million, the next a quarter. Programs cannot be planned around such wildly fluctuating budgets. Nongame administrators cannot be sure that programs begun one year will make it through the next. They cannot hire biologists with any assurance that the biologists' jobs will survive more than a year. It is thus a touch-and-go existence.

Another problem is the bandwagon effect. Once the success of nongame checkoffs became generally known, other special-interest groups horned in. The nongame programs were barely off the ground before other checkoffs started appearing on many tax forms. Taxpayers can be expected to respond only to a certain number of checkoffs, and competition from other quarters has diluted the contributions made to nongame. One study concluded that additional checkoffs will cut nongame donations by 70 percent.

Clearly, state nongame programs need a consistent source of funds. One of the best could come from state sales taxes. In 1976 voters in Missouri raised their sales tax one-eighth of 1 percent so money could be provided for wildlife conservation. Funds raised hover around $50 million a year. Because the money comes from taxpayers, not the legislature, the Missouri Department of Conservation is free to pursue wildlife management unencumbered by the whims of legislators.

So far, no states have taken up the Missouri plan. No sales-tax-for-wildlife bandwagon has been orchestrated. This suggests that the best way to fund nongame programs consistently is through the federal government. So far, no dice. Congress created a federal nongame program when it passed the Fish and Wildlife Conservation Act in 1980. The bill says that up to $5 million can be appropriated for the program each year. It allows the federal government to provide state wildlife agencies with 75 percent of the funds needed for nongame projects. Unfortunately, Congress has never provided any money to the program. The law is on the books, the cash is not. Financing, whether at state or federal levels, continues to be the hobgoblin of nongame programs.

Social problems have also plagued nongame endeavors. At first, many nonhunters saw the checkoffs as a chance to help pay for wildlife programs. Nonhunters are people who do not hunt but who don't care one way or another whether anyone else does. Of course, they were not the only segment of the population attracted to the checkoffs. They were joined by, among others, the anti-hunters.

Anti-hunters are the hardcore opponents of hunting. A survey of citizen knowledge about wildlife, conducted by Stephen Kellert of the Yale School of Forestry and Environmental Studies, has shown that anti-hunters and hunters are equals in their lack of general knowledge about wildlife. The overall populace tends to have a broader understanding of the natural community than either. Anti-hunters' concerns are mostly emotional. They do not like to see animals die. They feel bad about the loss of each individual animal. Their attitude differs from that of the wildlife manager or the professional conservationist, who are concerned with the survival of wildlife populations. The deaths of individual members of the population are no concern as long as they do not jeopardize the whole population. Obviously, conservationists who favor this approach have nothing against hunting that is well managed. They may not care too much for hunters, who have so frequently pushed wildlife biologists away from sound management, but they have nothing against their sport. Anti-hunters are diametrically opposed to this approach.

The anti-hunters are ardent in their opposition to hunting. They want nothing to do with it, and they do not want anyone else to have anything to do with it either. They might be expected to be strong supporters of nongame programs.

For a while it seemed they were. Then the anti-hunters took note of something that seemed amiss in the nongame programs. Some states were using nongame funds to buy land that was to be used for nongame protection. And then these states were turning around and opening the areas to hunting. Nongame funds, limited as they were, were being used to meet hunter needs. The anti-hunters bolted, and the nongame programs became subject to a vituperative debate that continues today. Many anti-hunters have even tried to scuttle the nongame programs on the grounds that they are a bogus front for raising money for game management. Anti–hunters based this argument on instances in which nongame funds have been used, for example, to improve wetlands on waterfowl hunting areas. However, the argument is not entirely convincing. Work undertaken on a waterfowl management area in many cases may benefit nongame birds that also occur there.

One other problem—a slight one—has insinuated itself in the planning of nongame programs. When these programs were first initiated, many of us thought they would broaden wildlife management to cover all species. We already had game management and endangered species protection. Now we would manage nongame animals. By studying them we would learn more about the relationships between species. We would learn more about how animals and plants interact with their environment. And we would help ensure that nongame species did not turn into endangered species.

Ironically, I noticed when selecting species to cover in this book that most nongame programs concentrate on endangered or listed species. Among the most popular projects funded by nongame programs are those designed to bolster bald eagle and peregrine falcon populations. This is done by transferring eagles and peregrines, both federally listed under the Endangered Species Act, from areas where they occur in stable numbers to areas where they have been wiped out. Thanks to these programs, bald eagles and peregrines are nesting in areas from which they were only recently extirpated. Other nongame projects cover endangered bats or listed salamanders and so on. The emphasis seems to be on endangered and threatened species. It was hard to put together a group of chapters on nongame projects that involved nonendangered species.

Far be it from me to denigrate any project designed to reestablish bald eagle and peregrine populations. But I cannot help wondering whether these projects shouldn't be funded with endangered species money. Aren't endangered and threatened wildlife already, at least in theory, protected, studied, and managed under endangered species programs? Does the emphasis that nongame programs put on listed species suggest that the programs are falling short of their avowed goal of protecting species long ignored by wildlife managers?

Nongame programs offer a promise that wildlife management will soon encompass all species. So far, however, they have not yet taken us very far beyond the borders already established by game laws and endangered species conservation.

Epilogue:
An Age of
Destruction

It would be hard to imagine a darker and more discouraging time for a conservationist than the end of the nineteenth century. Wildlife was everywhere in shambles. Naturalists who had witnessed the demise of so many species fully expected to see the extinction of many more. Deer, bison, wild turkeys, pronghorns, waterfowl, elk, bighorn sheep, mountain goats, and others seemed ready to follow the passenger pigeon into the abyss. The whole vast array had been driven to the edge by the gun and the ax, by uncontrolled hunting and unbridled development. Laws for wildlife protection were weak and ignored. Little social precedent existed for the preservation of wild things. Knowledge of wildlife and its needs was rudimentary. These things considered, early conservationists had little to hope for. All they could do was struggle against all the portents and attempt to build from scratch a legal and social edifice that would encompass wildlife protection. In the face of overwhelming odds they had to persevere. It is no wonder that wildlife literature at the turn of this century is studded with illusions to the anticipated loss of deer and other wildlife.

Now, as another century nears its end, the whole complexion of wildlife conservation has changed. More has been discovered about wildlife during the past two decades than was ever known before. If knowledge is power, then society has unprecedented strength with which to protect nature and wild things. At the moment, deer, pronghorn, elk, bison, and the rest abound. Their ancient numbers are gone, but the animals are in no apparent danger of extinction. Not all creatures have rebounded so suc-

217

cessfully, but massive conservation efforts have been mounted to help save many of the heavily beleaguered. That any are being protected is a resounding change from what obtained a century ago. Society, government, and law have changed drastically in the past century. Wildlife is no longer a commodity to be used up without thinking. And yet, despite all this, the natural world is poised on the edge of a disaster so vast that it would stop even late–nineteenth-century conservationists in their tracks. Everything that was saved, protected, and preserved may soon be lost in a time so short that history will one day look back upon this period as the mere blinking of an eye.

The signs of this looming catastrophe are apparent everywhere. They appear in the rainshowers that are turning lakes and ponds throughout the nation and the world into acidic pools incapable of supporting life. They appear in grazing practices that have turned more than 10,000 square miles of U.S. rangeland into desert. They appear in the cutting of tropical rain forests, which threatens the survival of at least a million species. They appear in the loss of wildlife habitat to suburban development. They appear throughout the nation in underground water supplies that have become polluted by agricultural chemicals that have seeped into the earth. They appear at the edges of the sea, where ocean currents laden with human wastes have sullied beaches once used for recreation but now more often closed as threats to health. They appear in the pollution of air above the Arctic Circle, where oil development at Prudhoe Bay releases as many nitrogen oxide pollutants as Washington, D.C., where discharges from Europe and the Soviet Union have created in some areas levels of pollution that rival the worst levels on the East Coast. They appear around Green Bay, Wisconsin, labeled by one environmental consultant a toxic soup of industrial wastes. There shorebirds are plagued increasingly by birth defects and women are experiencing unusually high rates of miscarriage. They appear on Staten Island, a borough of New York City located downwind from New Jersey industrial complexes. There, according to a preliminary congressional report, cancer rates are unusually high, a result perhaps of wind-borne toxins from New Jersey.

Each sign suggests that the planet is overstressed. There may not be room enough on Earth for both humans and human development. The natural world is being destroyed so rapidly that scarcely any effort to protect it can keep up. Changes are occurring on an almost inconceivable scale. Grasslands, forests, rivers, streams, the entire atmosphere—nothing has been left unharmed. A close look at just one of several far-reaching threats, global warming, shows the tremendous danger apparently in the offing. If current trends in the destruction of tropical forests and the pollution of the atmosphere do not abate, and perhaps even if they do, the planet will warm by 3 degrees Celsius over the next half century. The world will change drastically. Animals and plants in the United States probably will have to shift their ranges north by 200 miles to accommodate the higher temperature. Under natural conditions of global warming they would have millennia to do it. They need this much time, since many species move slowly. Even some plants that produce light, air-borne seeds can shift only about 15 miles in a century. Human-caused warming gives them only half that time to go fifteen times the distance. Northern forests may shrink by nearly half. Rainfall declines in the Great Plains will destroy many grassland plants. Animals dependent on flora lost to global warming will vanish, too. People can perhaps adapt to loss of the forests and the crop-growing regions of the Great Plains, but wildlife probably cannot. Already the rate of extinction has escalated to four hundred times what it was in recent geologic periods.

Conservation is thus facing its greatest challenge. Laws to protect wildlife from gunfire are no longer enough. The tools that destroy wildlife today are chain saws and bulldozers, and the destruction is being wrought on an international scale. Entire species of plants and animals in Latin America, Africa, and Southeast Asia are falling to dam construction, land clearing, and housing development, projects funded at least in part with U.S. dollars.

What will be lost to human society is immeasurable, but enough is known about the natural world to provide some hints. For example, only three plant species supply half the world's food, but some seven thousand species are known to be edible. Tens of thousands more await discovery but may be wiped out before they are found. Fourteen hundred plant species in tropical forests may provide cures for cancer. One plant, the rosy periwinkle, provides a drug that raised the remission rate of childhood leukemia to 95 percent. How many such plant species will be lost to deforestation, and how many have already been lost, cannot be guessed accurately. We know only enough to make our ignorance seem frightening, because many living things seemingly without purpose are proving valuable to humanity. While many developers denigrate people who want to save the last individuals of a vanishing butterfly species or a rare freshwater mollusk, medical researchers are discovering that sea squirts—obscure little globs of marine life—contain a chemical that may stop certain types of cancer. Should we disregard whole species because today we see no use for them? Only a few years ago the Pacific yew was treated by lumbermen as a trash tree. Now its bark, too, is being used in cancer research. To lose any species on the grounds that it has no value is foolish. One of conservation's greatest challenges is to convince politicians and developers—bright, educated people—that not what they know, but what they do not know, about living things should guide them in creating measures for the protection of wildlife. A poorly studied species may have uses not yet imagined or meet needs not yet conceived.

The conservation movement is at heart an emotional one. It is predicated on the idea that it would have been great fun to have seen endless herds of bison thundering over the plains, a great thrill to have watched flocks of passenger pigeons filling the skies for days at a time. But it is founded on sound practical logic. The centerpiece of that logic is the idea that if the natural community is so thoroughly disrupted that vast numbers of species disappear, then all living things are at risk. When the dinosaurs died off, for example, they did not die off alone. Nearly two-thirds of all marine species went with them. If the trees of the tropical rain forest are wiped out, will they take us with them?

It is quite likely, of course, that they will, joined in doing so by the streams and air we have polluted and by the grasslands we have turned to deserts. All living things are interrelated, each eventually dependent for survival upon the health and integrity of the rest. No living thing—not even humanity—can be separated from all others, no living thing has a fate all its own.

We seem to be running out of the essentials of life, such as clean water and air and even untouched forests, just as in the nineteenth century we were running out of wildlife. A hundred years ago we changed our attitudes and stopped thinking about wildlife as an inexhaustible commodity to be used up heedlessly. We came to see it as a valuable and aesthetic resource to be nurtured and conserved. Now we need to change our basic values once again, need to learn to care for and nurture the land and the seas and the air we breathe. We cannot hope to save wildlife, nor ensure our own survival, while we go about destroying the Earth. The future thus depends on humanity coming finally to see that it is a part of the natural world—as vulnerable to destruction as any wolf or whale or clear running stream, as integral to the proper workings of the planet as the blowing of the winds and the shining of the sun.

Bibliography

Deer

America's Vanishing Rain Forest: A Report on Federal Timber Management in Southeast Alaska. 1986. Washington, D.C.: The Wilderness Society.

Carey, John. 1987. "Trouble in paradise." *National Wildlife* 25: 42–45.

Cole, John N. 1984. "Game management: declining deer herd jeopardizes Maine's wildlife agency." *Audubon* 86 (Jan.):104–05.

Cone, Joe, and Roger DiSilvestro. 1983. "Trouble in the Tongass." *Defenders* 58 (Jan./Feb.):2–8.

Downing, Robert. 1987. "Success story: white-tailed deer." In *Restoring America's Wildlife*, edited by Harmon Kallman. Washington, D.C.: Department of the Interior.

Flader, Susan L. 1974. *Thinking Like a Mountain: Aldo Leopold and the Evolution of an Ecological Attitude Toward Deer, Wolves, and Forests.* Columbia: Univeristy of Missouri Press.

Garrott, Robert A., *et al.* 1987. "Movements of female mule deer in northwest Colorado." *Journal of Wildlife Management* 51:634–43.

Halls, Lowell K. 1978. "White-tailed deer." In *Big Game of North America: Ecology and Management*, edited by John L. Schmidt and Douglas L. Gilbert. Harrisburg, Pa.: Stackpole Books.

———. 1984. *White-tailed Deer: Ecology and Management.* Harrisburg, Pa.: Stackpole Books.

Hoopes, David T. 1982. "Old-growth timber and wildlife management in Southeast Alaska: a question of balance." *Proceedings of the Forty-Seventh North American Wildlife Conference*:588–604.

Kessler, Winifred B. 1982. *Wildlife and Second-Growth Forests of Southeast Alaska: Problems and Potential for Management.* Juneau, Alas.: U.S. Forest Service in cooperation with the University of Idaho.

Laycock, George. 1987. "Trashing the Tongass." *Audubon* 89 (Nov.):110–27.

Leopold, Aldo. 1943. "Deer irruptions." *Transactions of the Wisconsin Academy of Sciences, Arts and Letters* 35:351-66.

Livingston, Stephanie D. 1987. "Prehistoric biogeography of white-tailed deer in Washington and Oregon." *Journal of Wildlife Managment* 51:649-54.

Loft, Eric R., *et al.* 1987. "Influence of cattle stocking rate on the structural profile of deer hiding cover." *Journal of Wildlife Management* 51:655-64.

Longhurst, William M., and W. Leslie Robinette. 1981. *Effects of Clearcutting and Timber Management on Sitka Black-Tailed Deer: A Report to the Forest Service.* Juneau, Alas.: U.S. Forest Service.

Mackie, Richard J. 1987. "Mule deer." In *Restoring America's Wildlife*, edited by Harmon Kallman. Washington, D.C.: Department of the Interior.

Madson, Chris. 1986. "To feed or not to feed." *Audubon* 88 (Mar.): 22-27.

Matthews, John W., and Donald E. McKnight. 1982. "Renewable resource commitments and conflicts in Southeast Alaska." *Proceedings of the Forty-Seventh North American Wildlife Conference*:573-82.

Mooty, Jack J. 1987. "Habitat use and seasonal range size of white-tailed deer in north central Minnesota." *Journal of Wildlife Management* 51:644-48.

Schoen, John W., Olof C. Wallmo, and Matthew D. Kirchhoff. 1981. "Wildlife-forest relationships: is a reevaluation of old growth necessary?" *Proceedings of the Forty-Sixth North American Wildlife Conference*:531-44.

Schorger, A.W. 1953. "The white-tailed in early Wisconsin." *Transactions of the Wisconsin Academy of Sciences, Arts and Letters* 42:197-247.

Seton, Ernest Thompson. 1953. *Lives of Game Animals.* Boston: Charles T. Bradford Co.

Taylor, Walter P. 1956. *The Deer of North America: The White-tailed, Mule and Black-tailed deer, Genus* Odocoileus. Harrisburg, Pa: The Stackpole Company; Washington, D.C.: The Wildlife Management Institute.

Wallmo, Olof C. 1981. *Mule and Black-tailed Deer of North America.* Lincoln: University of Nebraska Press.

Wallmo, O.C., and J. W. Schoen, eds. 1979. *Sitka Black-tailed Deer: Proceedings of a Conference in Juneau, Alaska.* Juneau, Alas.: U.S. Forest Service in cooperation with the Alaska Department of Fish and Game.

Wallmo, Olof C., and John W. Schoen. 1980. "Response of deer to secondary forest succession in Southeast Alaska." *Forest Science* 26: 448–62.

Turkey

Bakeless, John. 1950. *The Eyes of Discovery.* Philadelphia: J.P. Lippincott Co.

Grinnell, G.B. 1910. *American Game Bird Shooting.* New York: Forest and Stream Publishing Co.

Hewitt, Oliver H., ed. 1967. *The Wild Turkey and Its Management.* Washington, D.C.: The Wildlife Society.

Kurzejeski, Eric W., Larry D. Vangilder, and John B. Lewis. 1987. "Survival of wild turkey hens in Northern Missouri." *Journal of Wildlife Management* 51:188–93.

Lewis, John B. 1987. "Success story: the wild turkey" in *Restoring America's Wildlife,* edited by Harmon Kallman. Washington, D.C: U.S. Department of the Interior, Fish and Wildlife Service.

Mosby, Henry S., and Charles O. Handley. 1943. *The Wild Turkey in Virginia: Its Status, Life History, and Management.* Richmond, Va.: P–R Projects, Division of Game and the Commission of Game and Inland Fisheries.

Ranson, Dean Jr., Orvin J. Rongstoel, and Donald H. Rusch. 1987. "Nesting ecology of Rio Grande turkeys." *Journal of Wildlife Management* 51:435–39.

Williams, Lovett Jr. 1981. *The Book of the Wild Turkey.* Tulsa, Okla.: Winchester Press.

Pronghorn

Barton, Katherine. 1987. "Bureau of Land Management." In *Audobon Wildlife Report 1987,* edited by Roger L. DiSilvestro. New York: National Audubon Society.

Cadieux, Charles L. 1987. "Pronghorn management: great plains rebound." In *Restoring America's Wildlife*, edited by Harmon Kallman. Washington D.C.: U.S. Department of the Interior, Fish and Wildlife Service.

Einarsen, Arthur S. 1948. *The Pronghorn Antelope and Its Management*. Washington, D.C.: Wildlife Management Institute.

Hornaday, William T. 1913. *Our Vanishing Wildlife*. New York: Charles Scribner and Sons.

Knipe, Theodore. 1944. *The Status of the Antelope Herds of Northern Arizona*. Phoenix: Arizona Game and Fish Commission.

Ligon, J. Stokley. 1946. *History and Management of Merriam's Wild Turkey*. Albuquerque: University of New Mexico Press.

Seton, Ernest Thompson. 1953. *Lives of Game Animals*. Boston: Charles T. Branford Co.

Skinner, M.P. 1924. *The American Antelope in Yellowstone National Park*. Syracuse, N.Y.: Roosevelt Wild Life Forest Experiment Station.

Yoakum, James D. 1978. "Pronghorn." In *Big Game of North America: Ecology and Management*, edited by John L. Schmidt and Douglas L. Gilbert. Harrisburg, Pa.: Stackpole Books.

————. 1988. "The Pronghorn." In *Audubon Wildlife Report 1988*, edited by William J. Chandler. San Diego, Cal.: Academic Press.

Waterfowl

Bellrose, Frank C. 1944. "Duck populations and kills; an evaluation of some waterfowl regulations in Illinois." *Bulletin of the Illinois Natural History Survey*. Urbana: Department of Registration and Education.

Butcher, Gregory S. 1987. "Data that count." *The Living Bird Quarterly* 6 (Autumn):20–23.

Chandler, William J. 1985. "Migratory Bird Protection and Management." In *Audubon Wildlife Report 1985*, edited by Roger L. DiSilvestro. New York: National Audubon Society.

————. 1986. "Migratory Bird Protection and Management." In *Audubon Wildlife Report 1986*, edited by Roger L. DiSilvestro. New York: National Audubon Society.

————. 1987. "Migratory Bird Protection and Management." In *Audubon Wildlife Report 1987*, edited by Roger L. DiSilvestro. San Diego, Cal.: Academic Press.

Grinnell, George Bird. 1901. *American Duck Shooting.* New York: Forest and Stream Publishing Co.

Hansen, Paul W. 1987. *Acid Rain and Wildlife: The Case for Concern in North America.* Arlington, Va.: Izaak Walton League of America.

Hawkins, A.S., R.C. Hanson, H.K. Nelson, and H.M. Reeves, eds.1987. *Flyways: Pioneering Waterfowl Management in North America.* Washington, D.C.: U.S. Department of the Interior, Fish and Wildlife Service.

Husar, John. 1988. "Battle in Mississippi still raging over US draining of bottomlands." *Chicago Tribune.* 13 March.

Krementz, David G., Michael J. Conroy, James E. Hines, and H. Franklin Percival. 1988. "The effects of hunting on survival rates of American black ducks." *Journal of Wildlife Management* 52:214–26.

Leffingwell, William Brace. 1890. *Shooting on Upland, Marsh, and Stream.* Chicago: Rand McNally and Co.

Luoma, Jon R. 1985. "Twilight in pothole country." *Audubon* 87: 66–85.

Mumford, Russell E. 1954. *Waterfowl Management in Indiana.* Indianapolis: Indiana Department of Conservation.

Ratti, John T., Lester D. Flake, and W. Alan Wentz, eds. 1982. *Waterfowl Ecology and Management: Selected Readings.* Bethesda, Md.: The Wildlife Society.

Sidle, John G., and Keith W. Harman. 1987. "Prairie pothole politics." *Wildlife Society Bulletin* 15:355–62.

Steinhart, Peter. 1987. "Empty the skies." *Audubon* 89 (Nov.):70–97.

Sullivan, Cheryl. 1987. "California refuge cleanup is key to wildfowl poisoning problem." *Christian Science Monitor.* 1 Dec. p. 5.

Tober, James A. 1981. *Who Owns the Wildlife: The Political Economy of Conservation in Nineteenth Century America.* Westport, Conn.: Greenwood Press.

Wolf

Aderhold, Mike. 1987. "All for the wolf." *Montana Outdoors* 18 (Sept./Oct.):15–22.

Antypas, Alex. 1988. "Old fears and superstition block return of Mexican wolf." *Audubon Activist* 2 (Jan./Feb.):5.

Brown, David E. 1983. *The Wolf in the Southwest.* Tucson: The University of Arizona Press.

Fischer, Hank. 1987. "Deep freeze for wolf recovery?" *Defenders* 62: 28–33.

Harbo, Samuel J., Jr., and Frederick C. Dean. 1983. "Historical and current perspectives on wolf management in Alaska." In *Wolves in Canada and Alaska*, edited by Ludwig N. Carbyn. Ottawa: Canadian Wildlife Service.

Miller, Thomas. 1988. "Wyoming wolves? Yes!" *Wyoming Wildlife* (March):4–7.

Peterson, Rolf L., Theodore N. Bailey, and James D. Woolington. 1983. "Wolf management and harvest patterns on the Kenai National Wildlife Refuge, Alaska." In *Wolves in Canada and Alaska*, edited by Ludwig N. Carbyn. Ottawa: Canadian Wildlife Service.

———. 1986. "The wolf." In *Audubon Wildlife Report 1986*, edited by Roger L. DiSilvestro. New York: National Audubon Society.

Ream, Robert R., Michael W. Fairchild, Diane A. Boyd, Daniel H. Pletscher. 1987. *Wolf Monitoring and Research in and Adjacent to Glacier National Park.* Missoula: School of Forestry and Montana Cooperative Wildlife Research Unit, University of Montana.

Seton, Ernest Thompson. 1953. *The Lives of Game Animals.* Boston: Charles T. Branford Co.

Simpson, Alan. 1988. "Wyoming wolves? No!" *Wyoming Wildlife* (March):8–11.

Tilt, Whitney, Ruth Norris, and Amos S. Eno. 1987. *Wolf Recovery in the Northern Rocky Mountains.* New York: National Audubon Society; Washington, D.C.: National Fish and Wildlife Foundation.

Townsend, Barbara. 1987. *Wolf: Annual Report of Survey-Inventory Activities.* Juneau, Alas.: Alaska Department of Fish and Ge ie.

Weise, Thomas F., William L. Robinson, Richard A. Ho к, and L. David Mech. 1975. *An Experimental Translocation of the Eastern Timber Wolf.* New York: National Audubon Society; Washington, D.C.: U.S. Fish and Wildlife Service.

Young, Stanley. 1946. *The Wolf in North American History.* Caldwell, Id.: The Caxton Printers, Ltd.

———, and Edward A. Goldman. 1944. *The Wolves of North America.* New York: Dover Books.

Grizzly Bear

Brown, David E. 1985. *The Grizzly in the Southwest*. Norman: University of Oklahoma Press.

DeVoto, Bernard, ed. 1953. *The Journals of Lewis and Clark*. Boston: Houghton Mifflin Company.

McCracken, Harold. 1955. *The Beast That Walks Like a Man*. Garden City, N.Y.: Hanover House.

McNamee, Thomas. 1984. *The Grizzly Bear*. New York: McGraw-Hill Book Company.

Robbins, Jim. 1988. "When species collide." *National Wildlife* 26 (Feb./Mar.):20–27.

Servheen, Chris. 1985. "The Grizzly Bear." In *Audubon Wildlife Report 1985*, edited by Roger L. DiSilvestro. New York: National Audubon Society.

Seton, Ernest Thompson. 1953. *Lives of Game Animals*. Boston: Charles T. Branford Company.

Shepard, Paul, and Barry Sanders. 1985. *The Sacred Paw: The Bear in Nature and Myth*. New York: Viking Penguin.

Condor

Harris, Harry. 1941. "The annals of *Gymnogyps* to 1900." *The Condor* 43:3–55.

Kiff, Lloyd. 1983. "A historical perspective on the condor." *Outdoor California* 44:5.

Koford, Carl B. 1953. *The California Condor*. New York: National Audubon Society.

Mallette, Robert D. 1966. "First cooperative study of the California condor." *California Fish and Game* 52:185–203.

Miller, Alden H., Ian I. McMillan, and Eben McMillan. 1965. *The Current Status and Welfare of the California Condor*. New York: National Audubon Society.

Ricklefs, Robert E., ed. 1978. *Report of the Advisory Panel on the California Condor*. New York: National Audubon Society.

Townsend, J.K. 1839. *Ornithology in the United States*. Philadelphia: J.B. Chevalier.

230   THE ENDANGERED KINGDOM

Wilbur, Sanford R. 1978. *The California Condor, 1966–76: A Look At Its Past and Future.* Washington, D.C.: Department of the Interior, Fish and Wildlife Service.

Bowhead Whale

Alaska Outer Continental Shelf Beaufort Sea Sale 97: Final Environmental Impact Statement. Vol. 1. 1987. Washington, D.C.: U.S. Department of the Interior.

Annual Report of the Marine Mammal Commission, Calendar Year 1987: A Report to Congress. 1988. Washington, D.C.: Marine Mammals Commission.

Ellis, Richard. 1981. *The Book of Whales.* New York: Alfred A. Knopf.

Fraker, Mark A. 1984. *Balaena mysticetus: Whales, Oil, and Whaling in the Arctic.* Anchorage, Alas.: Sohio Alaska Petroleum Company and BP Alaska Exploration, Inc.

McVay, Scott. 1973. "Stalking the Arctic Whale." *Scientific American* 61:24–37.

Minasian, Stanley M., Kenneth C. Balcomb III, and Larry Foster. 1984. *The World's Whales.* Washington, D.C.: Smithsonian Books.

Whipple, A.B.C., and the Editors of Time-Life Books. 1979. *The Whalers.* Alexandria, Va.: Time-Life Books.

Western Diamondback Rattlesnake

Klauber, Laurence M. 1982. Abridged by Karen Marvey McClung. *Rattlesnakes, Their Habits, Life Histories, and Influence on Mankind.* Berkeley: University of California Press.

River Otters

Chanin, Paul. 1985. *The Natural History of Otters.* New York: Facts on File.

Gilbert, Bil. 1982. "The utterly delightful otter." *Sports Illustrated* (13 Dec.):73–86.

Greiss, Jane M. 1987. "River otter reintroduction in Great Smoky Mountains National Park." Unpublished master's thesis for the University of Tennessee.

Johnson, Kyle. 1981. "Man-related factors tied to river otter deaths." University of Idaho, Moscow, news release for 24 June 1981.

Polechla, Paul J. Jr. 1988. "The nearctic river otter." In *Audubon Wildlife Report 1988/1989*, edited by William J. Chandler. San Diego, Cal.: Academic Press.

Schorger, A.W. 1951. "A brief history of the steel trap and its use in North America." *Transactions of the Wisconsin Academy of Sciences, Arts and Letters* 40:171–99.

Seton, Ernest Thompson. 1953. *Lives of Game Animals*. Boston: Charles T. Branford Co.

Bats

Ackerman, Diane. 1988. "Bats." *The New Yorker* (Feb.):37–62.

Anonymous. 1985. "Mines and bats." *Bats: Newsletter of Bat Conservation International* 2 (Feb.):3.

———. 1986. "States focus on bat conservation." *Bats: Newsletter of Bat Conservation International* 3 (Sept.):3.

———. 1987. "Vandals destroy hibernating Indiana bats." *Bats* 5 (June):5, 8.

———. 1988. "Long-nosed bats proposed for endangered status." *Bats* 6 (Summer):2, 16.

———. 1988. "Endangered species recovery team reports gains and losses." 6 (Summer):2, 16.

———. 1988. "Cave resources protection act needs your help." 6 (Summer):2, 16.

Blodget, Bradford G. 1984. "Tax check-off funds Massachusetts nongame and endangered species program." *Endangered Species Technological Bulletin* 9:4–7.

Clawson, Richard L. 1987. "Indiana bats: down for the count." *Bats* 5 (June):3–4.

McCracken, Gary F. 1986. "Why are we losing our Mexican free-tailed bats?" *Bats: Newsletter of Bat Conservation International* 3 (Sept.): 1–2, 4.

———. 1988. "Red lists, green lists, and bat conservation." In press for *Bats* (Fall):6.

Perkins, Mark. 1985. "The plight of *Plecotus.*" *Bats: Newsletter of Bat Conservation International* 2 (February):1–2.

Tuttle, Merlin D. 1981. *Bats and Public Health.* Milwaukee: Milwaukee Public Museum Press.

———. 1985. "Joint effort saves vital bat cave." *Bats: Newsletter of Bat Conservation International* 2 (Dec.):3.

———. 1986. "Endangered gray bat benefits from protection." *Bats: Newsletter of Bat Conservation International* 4 (December):1–2.

———. 1988. *The Importance of Bats.* Austin, Texas: Bat Conservation International.

Migratory Birds

Hamel, Paul B. 1988. "Bachman's Warbler." In *Audubon Wildlife Report 1988/89.* San Diego, Cal.: Academic Press.

Myers, Norman. 1984. *The Primary Source.* New York: W.W. Norton & Company.

Powell, George V.N., and John H. Rappole. 1986. "The Hooded Warbler." In *Audubon Wildlife Report 1986*, edited by Roger L. DiSilvestro. New York: National Audubon Society.

Rappole, John H., Eugene S. Morton, Thomas E. Lovejoy III, James L. Ruos, and Byron Swift. 1983. *The Nearctic Avian Migrants in the Neotropics.* Washington, D.C.: U.S. Department of the Interior.

Wilson, E.O. 1988. *Biodiversity.* Washington, D.C.: National Academy Press.

Index

333.95
Di

DiSilvestro, Roger
Endangered Kingdom

$10.95

DATE DUE			
Jan 10			